Foundations of Statistical Inference

Contributions to Statistics

Yoel Haitovsky · Hans Rudolf Lerche
Yaacov Ritov (Editors)

Foundations
of Statistical Inference

Proceedings of the Shoresh Conference 2000

With 23 Figures and 18 Tables

Physica-Verlag

A Springer-Verlag Company

Series Editors
Werner A. Müller
Martina Bihn

Editors

Professor Dr. Yoel Haitovsky
Professor Dr. Yaacov Ritov
Department of Statistics
Hebrew University
91905 Jerusalem
Israel

Professor Dr. Hans Rudolf Lerche
Department of Mathematical Stochastics
University of Freiburg
Eckerstraße 1
79104 Freiburg
Germany

ISSN 1431-1968
ISBN 3-7908-0047-3 Physica-Verlag Heidelberg New York

Cataloging-in-Publication Data applied for
A catalog record for this book is available from the Library of Congress.
Bibliographic information published by Die Deutsche Bibliothek
Die Deutsche Bibliothek lists this publication in the Deutsche Nationalbibliografie; detailed bibliographic
data is available in the Internet at <http://dnb.ddb.de>.

Physica-Verlag Heidelberg New York
a member of BertelsmannSpringer Science+Business Media GmbH

© Physica-Verlag Heidelberg 2003
Printed in Germany

Softcover Design: Erich Kirchner, Heidelberg

SPIN 10864383   88/3130-5 4 3 2 1 0 – Printed on acid-free and non-aging paper

Preface

This volume is a collection of papers presented at a conference held in Shoresh Holiday Resort near Jerusalem, Israel, in December 2000 organized by the Israeli Ministry of Science, Culture and Sport. The theme of the conference was "Foundation of Statistical Inference: Applications in the Medical and Social Sciences and in Industry and the Interface of Computer Sciences". The following is a quotation from the Program and Abstract booklet of the conference. "Over the past several decades, the field of statistics has seen tremendous growth and development in theory and methodology. At the same time, the advent of computers has facilitated the use of modern statistics in all branches of science, making statistics even more interdisciplinary than in the past; statistics, thus, has become strongly rooted in all empirical research in the medical, social, and engineering sciences. The abundance of computer programs and the variety of methods available to users brought to light the critical issues of choosing models and, given a data set, the methods most suitable for its analysis. Mathematical statisticians have devoted a great deal of effort to studying the appropriateness of models for various types of data, and defining the conditions under which a particular method work."

In 1985 an international conference with a similar title* was held in Israel. It provided a platform for a formal debate between the two main schools of thought in Statistics, the Bayesian, and the Frequentists. Since that time an interesting development in the field has been the narrowing of the gap between the two approaches to Statistical Inference. The step towards reconciliation has been facilitated by a breakthrough in computing, that took place over the last fifteen years in both hardware and software which made it possible to perform long and complicated calculations. The availability of newly developed tools necessary for Bayesian applications has enabled the specification of more realistic models and the popularization of the Bayesian approach in applied studies. At the same time, the development in calculation has brought a flourishing of non-parametric methods, whose advantage over the classical methods is their ability to handle less rigid models. A main emphasis of the recent conference was to elaborate on these developments.

The volume surveys some aspects of the discussion. The papers are presented in four groups: Part I: Identification with Incomplete Observations, Data Mining, Part II: Bayesian Methods and Modelling, Part III: Testing,

---

* The econometric oriented papers of the 1985 conference were published by the Journal of Econometrics, Vol. 37, No. 1, 1988 entitled: *Competing Statistical Paradigms in Econometrics*, Teun Kloek and Yoel Haitovsky (eds).

Goodness of Fit and Randomness, Part IV: Statistics of Stationary Processes.

Sponsoring institutions of the conference in 2000 were US National Science Foundation, European Union, The Israeli Academy of Science and Humanities, The Hebrew University, FSTRS – Frontier Science and Technology Research Foundation, INTAS – International Association for the Promotion of Cooperation with Scientists from the former Soviet Union, Teva Pharmaceutical Industries, Israeli Central Bureau of Statistics, Israeli Statistical Association, The British Council.

Finally, the appearance of the volume is due to Ines Giers, Monika Hattenbach, and Thomas Lais, who did a masterful job of retyping some of the papers and of bringing the volume to uniform format.

Jerusalem,   *Yoel Haitovsky*
December 2002   *Hans R. Lerche*
                                                               Yaacov Ritov

Contents

Part III. Testing, Goodness of Fit and Randomness

**Asymptotic Expansions for Long-Memory
Stationary Gaussian Processes** 217
*David M. Zucker, Judith Rousseau, Anne Philippe and Offer
Lieberman*

Part I

Identification with Incomplete Observations, Data Mining

Bounding Entries in Multi-way Contingency Tables Given a Set of Marginal Totals

Adrian Dobra[1] and Stephen E. Fienberg[2]

[1] National Institute of Statistical Sciences, Research Triangle Park, North Carolina, USA
[2] Department of Statistics and Center for Automated Learning and Discovery, Carnegie Mellon University, Pittsburgh, USA

Abstract. We describe new results for sharp upper and lower bounds on the entries in multi-way tables of counts based on a set of released and possibly overlapping marginal tables. In particular, we present a generalized version of the shuttle algorithm proposed by Buzzigoli and Giusti that computes sharp integer bounds for an arbitrary set of fixed marginals. We also present two examples which illustrate the practical import of the bounds for assessing disclosure risk.

1 Introduction

In this paper, we provide an overview of our recent work to develop bounds for entries in contingency and other non-negative tables (see also [8]). Our interest in this problem grows out of work to develop a Web-based table query system, coordinated by the National Institute of Statistical Sciences in the spirit of a pilot system described by Keller-McNulty and Unger [19]. The system is being designed to work with a database consisting of a k-way contingency table and it allows only those queries that come in the form of requests for marginal tables. What is intuitively clear from statistical theory is that, as margins are released and cumulated by users, there is increasing information available about the table entries. The system must examine each new query in combination with those previously released margins and decide if the risk of disclosure of individuals in the full unreleased k-way table is too great. Then it might offer one of three responses: (1.) yes-release; (2.) no-don't release; or perhaps (3.) simulate a new table, which is consistent with the previously released margins, and then release the requested margin table from it (c.f. [9], [14], [15]).

There are various approaches to assessing risk of disclosure and most of them relate to the inadvertent "release" of small counts in the full k-way table (e.g. see [13], [23], [24]). Here we follow the approach of examining upper and lower bounds on the cell entries (see [3], [4], [12], [22]). For more general background on related methods of disclosure limitation, we refer the interested reader to [25], [26].

The approach we outline in this paper draws heavily on the ideas associated with the theory of log-linear models for contingency tables ([1], [20]), where the minimal sufficient statistics are in fact marginal totals corresponding to the highest-order terms in the model. In Section 2, we give some technical background and then, in Section 3, we present results from [7] corresponding to decomposable and reducible graphical models. Then, in Section 4, we outline a general algorithm that computes sharp bounds for margins corresponding to any standard log-linear model. This algorithm generalizes the "shuttle" algorithm approach suggested by Buzzigoli and Giusti [3]. We apply our results to two examples, a 2^6 table and a 2^{16} table, and we discuss some of the possible implications for disclosure.

2   Technical Background

Bounds for entries in two-way contingency tables go back to seminal papers by Bonferonni [2], Fréchet [16], and Hoeffding [17]. For an $I \times J$ table with entries $\{n_{ij}\}$ and row margins $\{n_{i+}\}$ and column margins $\{n_{+j}\}$, these bounds take the form

$$\min\{n_{i+}, n_{+j}\} \geq n_{ij} \geq \max\{0, n_{i+} + n_{+j} - n_{++}\}. \qquad (1)$$

For simplicity, we refer to these as *Fréchet bounds*. Until recently, the only multi-dimensional generalizations of this result that have been utilized involved non-overlapping fixed marginals. Our interest has been in deriving computationally efficient approaches to computing bounds when the marginals overlap (c.f. the related work described in Joe [18]).

Any contingency table with non-negative integer entries and fixed marginal totals is a lattice point in the convex polytope $\mathbf{Q}$ defined by the linear system of equations induced by the released marginals. The constraints given by the values in the released marginals induce upper and lower bounds on the interior cells of the initial table. These bounds or *feasibility intervals* can be obtained by solving the corresponding linear programming problems. The importance of systematically investigating these linear systems of equations should be readily apparent. If the number of lattice points in $\mathbf{Q}$ is below a certain threshold, we have significant evidence that a potential disclosure of the entire dataset might have occurred. Moreover, if the induced upper and lower bounds are too tight or too close to the actual sensitive value in a cell entry, the information associated with the individuals classified in that cell may become public knowledge.

The problem of determining sharp upper and lower bounds for the cell entries subject to some linear constraints expressed in this form is known to be NP-hard (see Roehrig et al. [22]). Several approaches have been proposed for computing bounds: however, almost all of them have drawbacks that show the need for alternate solutions. Network models (c.f. [4]) need formal structure to work even for 3-way tables and besides there is no general formulation for higher-way tables. In some ways, the most natural method for solving linear programming problems is the simplex method. For the bounds problem, we would have to run the procedure twice for every element in the table and consequently we overlook the underlying dependencies among the marginals by regarding the maximization/minimization problem associated with some cell as unrelated to the parallel problems associated with the remainder of the cells in the table. Although the simplex method works well for small problems and dimensions, by employing it we would ignore the special structure of the problem because we would consider every table as a linear list of cells. The computational inadequacy of the simplex approach is further augmented by the fact that we may get fractional bounds (see Cox [4]), which are very difficult to interpret. To avoid fractional bounds, one would have to make use of integer programming algorithms, but their computational complexity prevent their usage even for problems of modest size. These considerations suggest the need for more specialized, computationally inexpensive algorithms that could fully exploit the special structure of the problem we are dealing with.

3   Bounds when Marginals Characterize Decomposable and Reducible Graphical Models

We visualize the dependency patterns induced by the released marginals by constructing an independence graph for the variables in the underlying cross-classification. Each variable cross-classified in the table is associated with a vertex in this graph. If two variables are not connected, they are conditionally independent given the remainder. Models described solely in terms of such conditional independencies are said to be *graphical* (e.g. see Lauritzen [20]).

3.1   Bound Results

Decomposable graphical models have closed form structure and special properties. The expected cell values can be expressed as a function of the fixed marginals. To be more explicit, the maximum likelihood estimates are the product of the marginals divided by the product of the separators. By induction on the number of MSSs, in [7], we developed generalized Fréchet bounds for decomposable log-linear models with any number of MSSs. These generalized Fréchet bounds are sharp in the sense that they are the tightest possible bounds given the marginals. In addition, we can determine feasible tables for which these bounds are attained.

Theorem 1 (Fréchet Bounds for Decomposable Models)   *Assume that the released set of marginals for a k-way contingency table is the set of MSSs of a decomposable log-linear model. Then the upper bounds for the cell entries in the initial table are the minimum of relevant margins, while the lower bounds are the maximum of zero, or sum of the relevant margins minus the separators.*

When the log-linear model associated with the released set of marginals is not decomposable, it is natural to ask ourselves whether we could reduce the computational effort needed to determine the tightest bounds by employing the same strategy used for decomposable graphs, i.e. decompositions of graphs by means of complete separators. An independence graph that is not necessarily decomposable, but still admits a proper decomposition, is called *reducible* (Leimer [21]). Once again, we point out the link with maximum likelihood estimation in log-linear models. We define a *reducible log-linear model* in [7] as one for which the corresponding MSSs are marginals that characterize the components of a reducible independence graph. If we can calculate the maximum likelihood estimates for the log-linear models corresponding to every component of a reducible graph G, then we can easily derive explicit formulae for the maximum likelihood estimates in the reducible log-linear model with independence graph G [7].

Theorem 2 (Fréchet Bounds for Reducible Models) *Assume that the released set of marginals is the set of MSSs of a reducible log-linear model. Then the upper bounds for the cell entries in the initial table are the minimum of upper bounds of relevant components, while the lower bounds are the maximum of zero, or sum of the lower bounds of relevant components minus the separators.*

3.2   Example 1: Risk Factors for Czech Auto Workers

The data in Table 1 come from a prospective epidemiological study of 1841 workers in a Czechoslovakian car factory, as part of an investigation of potential risk factors for coronary thrombosis (see Edwards and Havranek [10]). In left-hand panel of Table 1, A indicates whether or not the worker "smokes", B corresponds to "strenuous mental work", C corresponds to "strenuous physical work", D corresponds to "systolic blood pressure", E corresponds to "ratio of β and α lipoproteins" and F represents "family anamnesis of coronary heart disease". Assume we are provided with three marginal tables [BF], [ABCE], and [ADE] of this 6-way table. These are the marginals corresponding to a graphical model whose independence graph is given in Fig. 1, and this model fits the data well.

Using the result from Theorem 1, we see that the upper bounds for the cell entries induced by the marginals [BF], [ABCE], and [ADE] are the minimum of the corresponding entries in the fixed marginals, while the lower bounds are the sum of the same entries minus the sum of the corresponding entries in the marginals associated with the separators of the independence graph, i.e., [B] and [AE]. We give these bounds in the right-hand panel of Table 1. There are three cell entries containing non-zero "small" counts, i.e. counts of "1" and "2" in Table 1. The corresponding bounds are [0,25], [0,38] and [0,20]. Since the latter two of these differ, we see that the upper and lower bounds are therefore dependent not only on the fixed marginals, but also on

Table 1. Czech autoworkers data from [10]. The left-hand panel contains the cell counts and the right-hand panel contains the bounds given the margins [BF], [ABCE], and [ADE]

				B	no	yes		B	no		yes	
F	E	D	C	A	no yes	no yes		A	no	yes	no	yes
neg	< 3	< 140	no		44 40	112 67			[0,88]	[0,62]	[0,224]	[0,117]
			yes		129 145	12 23			[0,261]	[0,246]	[0,25]	[0,38]
		≥ 140	no		35 12	80 33			[0,88]	[0,62]	[0,224]	[0,117]
			yes		109 67	7 9			[0,261]	[0,151]	[0,25]	[0,38]
	≥ 3	< 140	no		23 32	70 66			[0,58]	[0,60]	[0,170]	[0,148]
			yes		50 80	7 13			[0,115]	[0,173]	[0,20]	[0,36]
		≥ 140	no		24 25	73 57			[0,58]	[0,60]	[0,170]	[0,148]
			yes		51 63	7 16			[0,115]	[0,173]	[0,20]	[0,36]
pos	< 3	< 140	no		5 7	21 9			[0,88]	[0,62]	[0,126]	[0,117]
			yes		9 17	1 4			[0,134]	[0,134]	[0,25]	[0,38]
		≥ 140	no		4 3	11 8			[0,88]	[0,62]	[0,126]	[0,117]
			yes		14 17	5 2			[0,134]	[0,134]	[0,25]	[0,38]
	≥ 3	< 140	no		7 3	14 14			[0,58]	[0,60]	[0,126]	[0,126]
			yes		9 16	2 3			[0,115]	[0,134]	[0,20]	[0,36]
		≥ 140	no		4 0	13 11			[0,58]	[0,60]	[0,126]	[0,126]
			yes		5 14	4 4			[0,115]	[0,134]	[0,20]	[0,36]

the position they occupy in the cross-classification. Moreover, the bounds for the entry of "1" are wider than the bounds for one of the entries of "2". At any rate, all three of these pairs of bounds differ quite substantially and thus we might conclude that there is little chance of identifying the individuals in the small cells.

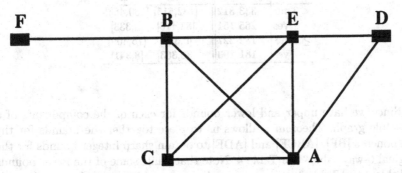

Fig. 1. Independence graph induced by the marginals [BF], [ABCE] and [ADE]

Now we step back and look at an even less problematic release involving the margins: [BF], [BC], [BE], [AB], [AC], [AE], [CE], [DE], [AD]. The independence graph associated with this set of marginals is the same graph in

Fig. 1 but the log-linear model whose MSSs correspond to those marginals is not graphical. Since the independence graph decomposes in three components, [BF], [ABCE], and [ADE], and two separators, [B] and [AE], as we have seen, we can apply the result from Theorem 2.

The first component, [BF], is assumed fixed; hence there is nothing to be done. The other two components are not fixed, however, and we need to compute upper and lower bounds for each of them. Using the algorithm presented in the next section, we calculated bounds for the cell entries in the marginal [ABCE] given the marginals [BC], [BE],[AB], [AC], [AE], [CE] (see Table 2). We did the same for the marginal [ADE] given the marginals [AE], [DE], [AD] (see Table 3).

Table 2. Marginal [ABCE] from Table 1 and bounds for this marginal given all 2-way totals

E	C	B no		yes		B no		yes	
		A no	yes	no	yes	A no	yes	no	yes
< 3	no	88	62	224	117	[0,206]	[0,167]	[0,404]	[0,312]
	yes	261	246	25	38	[0,421]	[30,463]	[0,119]	[0,119]
≥ 3	no	58	60	170	148	[0,181]	[0,167]	[0,363]	[0,339]
	yes	115	173	20	36	[0,314]	[0,344]	[0,119]	[0,119]

Table 3. Marginal [AED] from Table 1 and bounds for this marginal given all 2-way totals

E	D	A no	yes	A no	yes
< 3	no	333	312	[182,515]	[130,463]
	yes	265	151	[83,416]	[0,333]
≥ 3	no	182	227	[0,333]	[76,409]
	yes	181	190	[30,363]	[8,341]

Since we have upper and lower bounds for each of the components of a reducible graph, Theorem 2 allows us to piece together the bounds for the components [BF], [ABCE] and [ADE] to obtain sharp integer bounds for the original 6-way table - see Table 4. Note that while some of the lower bounds in Table 2 and Table 3 are non-zero, when we combine to produce the bounds in Table 4 the resulting lower bounds are all zero.

We emphasize that Theorem 2 is a sound technique for replacing the original problem, namely, computing bounds for a 6-way table, by two smaller ones, i.e., computing bounds for a 4-way and a 3-way table. The computational effort required for implementing and using Theorem 2 is minimal once bounds for the components are available, and thus exploiting it in this fashion could lead to appreciable computational savings.

Table 4. Bounds for Czech auto-workers data from Table 1 given the marginals [BF], [BC], [BE],[AB], [AC], [AE], [CE], [DE], [AD]

				B	no		yes	
F	E	D	C	A	no	yes	no	yes
neg	< 3	< 140	no		[0,206]	[0,167]	[0,404]	[0,312]
			yes		[0,421]	[0,463]	[0,119]	[0,119]
		≥ 140	no		[0,206]	[0,167]	[0,404]	[0,312]
			yes		[0,416]	[0,333]	[0,119]	[0,119]
	≥ 3	< 140	no		[0,181]	[0,167]	[0,333]	[0,339]
			yes		[0,314]	[0,344]	[0,119]	[0,119]
		≥ 140	no		[0,181]	[0,167]	[0,363]	[0,339]
			yes		[0,314]	[0,341]	[0,119]	[0,119]
pos	< 3	< 140	no		[0,134]	[0,134]	[0,126]	[0,126]
			yes		[0,134]	[0,134]	[0,119]	[0,119]
		≥ 140	no		[0,134]	[0,134]	[0,126]	[0,126]
			yes		[0,134]	[0,134]	[0,119]	[0,119]
	≥ 3	< 140	no		[0,134]	[0,134]	[0,126]	[0,126]
			yes		[0,134]	[0,134]	[0,119]	[0,119]
		≥ 140	no		[0,134]	[0,134]	[0,126]	[0,126]
			yes		[0,134]	[0,134]	[0,119]	[0,119]

4 A General Bounds Algorithm

In Section 3, we took advantage of the special structure of the conditional independencies "induced" among the variables cross-classified in a table of counts by the set of fixed marginals. However, if all $(k-1)$-way marginal tables are given, the corresponding independence graph is complete, hence there are no conditional independence relationships to exploit. Fienberg [12] noted that, if the table is dichotomous, the log-linear model of no kth-order interaction has only one degree of freedom and consequently the counts in any cell can be uniquely expressed as a function of one single fixed cell alone. By imposing the non-negativity constraints for every cell in our contingency table, we are then able to derive sharp upper and lower bounds. It turns out that dichotomous tables are the key to derive sharp bounds for a k-way table given an arbitrary set of fixed marginals.

4.1 Terminology and Notation

Let **T** denote the set of cells of all possible tables that could be formed by collapsing the original k-way table **n** if not only across variables, but also across categories. The elements in **T** are essentially blocks formed by joining table entries in **n**. If the set of cell entries in **n** that define a "super-cell" $t_1 \in \mathbf{T}$ is included in the set of cells defining another "super-cell" $t_2 \in \mathbf{T}$, we write $t_1 \prec t_2$. With this partial ordering, $(\mathbf{T}, \prec)$ has a maximal element, namely the grand total of **n** and several minimal elements, i.e., the cell entries in the initial table **n**. The grand total of **n** is maximal because all the cells in **n** "contribute" to it. On the other hand, a cell entry in **n** is minimal in **T** since any block of cells in **T** is constructed from one single cell in **n** or by joining at least two other blocks. One can represent **t** as a hierarchy of cells induced by the ordering "$\prec$", with the grand total at the top level and the cells in **n** at the bottom level of the hierarchy.

Consider three blocks of cells t_1, t_2, and t_3. If t_2 can be formed by joining t_1 and t_3, we write

$$t_1 \oplus t_3 = t_2. \tag{2}$$

The operator "$\oplus$" is equivalent to joining two blocks of cells in **T** to form a third block. The blocks to be joined have to be composed from the same categories in $(k-1)$ dimensions and they are also required not to share any categories in the remaining dimension. If either of these conditions does not hold, their union is not going to be a block of cells in **T**. Denote by $L(t)$ and $U(t)$ the current upper and lower bounds for the "super-cell" $t \in \mathbf{T}$. Let

$$L(\mathbf{T}) := \{L(t) \ : \ t \in \mathbf{T}\} \text{ and } U(\mathbf{T}) := \{U(t) \ : \ t \in \mathbf{T}\}. \tag{3}$$

$L(t)$ and $U(t)$ are the bounds arrays we are trying to determine. Every $t \in \mathbf{T}$ could have a value $V(t)$ assigned to it. If t corresponds to an entry in a fixed marginal, we actually "know" the value $V(t)$ of that entry, hence we set the current lower bound and the current upper bound of t to be the known value $V(t)$.

Let $\mathbf{T_0}$ be the set of cells in **T** for which the lower bound is currently equal to the upper bound. These are the cells that have a value assigned to them:

$$V(t) = L(t) = U(t) \Leftrightarrow t \in \mathbf{T_0}. \tag{4}$$

When the iterative procedure described below starts, $\mathbf{T_0}$ will contain only the cells in the fixed marginals. For the remaining cells in **T**, we could set $L(t)$ and $U(t)$ to be the bounds $L_0(t), U_0(t)$ induced by fixing the one-dimensional marginals of **n**. These bounds are looser than the bounds we are trying to calculate since it is reasonable to assume that the one-dimensional marginals can be obtained by collapsing the marginals we consider to be fixed. In addition, the log-linear model induced by the one-dimensional marginals is decomposable, hence $L_0(t)$ and $U_0(t)$ can be easily calculated by employing Theorem 1. The intervals $[L(t), U(t)], t \in \mathbf{T}$, are the initial feasibility intervals for the iterative procedure we will describe below.

As the algorithm progresses, the bounds for the cells in **T** are improved and more and more cells are added to $\mathbf{T_0}$. To be more precise, "improving" the bounds means decreasing the upper bounds and increasing the lower bounds. When the bounds associated with a cell t become equal, the cell is included in T_0 and is assigned a value $V(t) := L(t) = U(t)$. We are now able to state the bounds problem in a new equivalent form:

"Find sharp integer bounds for the cells in **T**
if the values of some cells $T_0 \subset T$ *are fixed."*

4.2 The Generalized Shuttle Algorithm

The fundamental idea behind the "shuttle" algorithm is that the upper and lower bounds for the cells in **T** are interlinked. Although Buzzigoli and Giusti [3] sketched this innovative idea for the 3-way table problem given the three 2-way marginals, they did not accurately identify and exploit the full hierarchical structure of the cells contained in the marginals of a frequency count table **n**. The method we outline here builds on their approach and sequentially improves the bounds for all the cells we are interested in until no further adjustment can be made.

As before, we assume that, for every cell $t \in T$, we know a valid lower bound $L(t)$ and a valid upper $U(t)$. With these notations, the initial set of fixed cells is

$$T_0 := \{t \in T : L(t) = U(t)\}. \tag{5}$$

For all the cells t in T_0, we assign a value $V(t) := L(t) = U(t)$. We let $Q = Q(T)$ denote the triplets of cells

$$Q(T) := \{(t_1, t_2, t_3) \in T \times T \times T : t_1 \oplus t_3 = t_2\}, \tag{6}$$

which represent the cell dependencies we are trying to satisfy. We sequentially go through all these dependencies and update the upper and lower bounds in the following way. Consider a triplet $(t_1, t_2, t_3) \in Q$. We have $t_1 \prec t_2$ and $t_3 \prec t_2$. If all three cells have fixed values, i.e. $t_1, t_2, t_3 \in T_0$, we check whether we came across an inconsistency. The procedure stops if

$$V(t_1) + V(t_3) \neq V(t_2). \tag{7}$$

Assume that $t_1, t_3 \in T_0$ and $t_2 \notin T_0$. Then t_2 can only take one value, namely $V(t_1) + V(t_3)$. If $V(t_1) + V(t_3) \notin [L(t_2), U(t_2)]$, we encountered an inconsistency and exit the procedure. Otherwise we set

$$V(t_2) = L(t_2) = U(t_2) := V(t_1) + V(t_3), \tag{8}$$

and include t_2 in the set T_0 of cells having a fixed value. Similarly, if $t_1, t_2 \in T_0$ and $t_3 \notin T_0, t_3$ can only be equal to $V(t_2) - V(t_1)$. If $V(t_2) - V(t_1) \notin [L(t_3), U(t_3)]$, we again discovered an inconsistency. If this is not true, we set

$$V(t_3) = L(t_3) = U(t_3) := V(t_2) - V(t_1) \text{ and } T_0 := T_0 \cup \{t_3\}. \tag{9}$$

In the case when $t_2, t_3 \in \mathbf{T}_0$ and $t_1 \notin \mathbf{T}_0$, we proceed in an analogous manner. Now we examine the situation when at least two of the cells t_1, t_2, t_3 do not have a fixed value. For each of the three cells not having a fixed value, we update its upper and lower bounds so that the new bounds satisfy the dependency $t_1 \oplus t_3 = t_2$. Suppose $t_1 \notin \mathbf{T}_0$. Then the updated bounds for t_1 will be

$$U(t_1) := \min\{U(t_1), U(t_2) - L(t_3)\} \quad \text{and} \tag{10}$$
$$L(t_1) := \max\{L(t_1), L(t_2) - U(t_3)\}.$$

If $t_3 \notin \mathbf{T}_0$, we update $L(t_3)$ and $U(t_3)$ in the same way. Finally, if $t_2 \notin \mathbf{T}_0$, we set

$$U(t_2) := \min\{U(t_2), U(t_1) + U(t_3)\} \quad \text{and} \tag{11}$$
$$L(t_2) := \max\{L(t_2), L(t_1) + L(t_3)\}.$$

After updating the bounds of some cell $t \in \mathbf{T}$, we check whether the new upper bound is equal to the new lower bound. If this is true, i.e. $L(t) = U(t)$, we include t in the list of cells having a fixed value:

$$\mathbf{T}_0 := \mathbf{T}_0 \cup \{t\}, \tag{12}$$

and set $V(t) := L(t) = U(t)$. We continue going through all the dependencies in Q until the upper bounds no longer decrease, the lower bounds no longer increase and no new cells are added to $\mathbf{T}_0$. The procedure will come to an end if and only if an inconsistency is detected or if the upper and lower bounds cannot be subsequently improved. Either one of these two events will eventually occur, hence the procedure we described stops after a finite number of steps.

Unfortunately, the bounds we end up with are not necessarily sharp, except in: (i) the decomposable case, and (ii) the case of a dichotomous k-way table with all $(k-1)$-way marginals fixed. To be more explicit, if the marginals we fix are the MSSs of a decomposable log-linear model, the bounds calculated by the generalized shuttle algorithm will coincide with the bounds obtained by making use of Theorem 1, whereas in case (ii), the generalized shuttle algorithm will successfully determine the best integer bounds by expressing any cell as a function of any other cell, and then imposing the non-negativity conditions on these constraints.

For the general k-way bounds problem with an arbitrary set of fixed marginals, we need to "correct" the bounds by constructing feasible integer tables for which those bounds are actually attained. We explore the space $\mathbf{Q}$ by repeatedly assigning values to the cells in the original table. We do not perform an exhaustive search of $\mathbf{Q}$ since we immediately adjust the upper and lower bounds for the remaining cells in $\mathbf{T}$ once we pick a value for a cell entry, and consequently the values we attempt to assign to a particular cell are chosen from the current feasibility interval associated with that entry. Additional technical details can be found in Dobra [5].

We note that each bound can be checked independently of any other bound, hence adjusting the bounds can be done in parallel on a multi-processor machine. The computation time could be further decreased by using the following artifice: once a feasible integer table containing a count equal to a bound for some cell entry is constructed, we check to see whether other upper or lower bounds can also be found in that table. This way, we will not have to attempt to construct another table for these bounds. This simple trick proves to be very efficient in the case of large sparse contingency tables.

4.3 Example 1 Revisited

We have already applied this general algorithm to the separable components of the 6-way Czech auto-worker data in Table 1, to get sharp bounds for a separable table. Here we note what happens in the other special case when no "correction" is required for feasible tables: when all 5-way margins are released. The space of tables Q in this case contains only two integer tables: the original table n itself and a second table whose entries are found by adding or subtracting one unit from the corresponding entries in n. Consequently, the feasibility intervals $[L(t), U(t)]$ for all the cells in n have length one. This means that releasing all 5-way margins could well compromise the confidentiality of the individuals corresponding to the entries containing counts of "1" and "2" and perhaps even the entries containing the count of "3".

4.4 Example 2: The National Long Term Care Survey

Our second example involves a 2^{16} contingency table n extracted from the "analytic" data file for National Long-Term Care Survey created by the Center of Demographic Studies at Duke University. Each dimension corresponds to a measure of disability defined by an activity of daily leaving, and the table contains information cross-classifying individuals aged 65 and above. This extract involves data pooled across four waves of a longitudinal survey, and it involves sample as opposed to population data. We henceforth act *as if* these were population data. For a detailed description of this extract see [11].

We have applied the generalized shuttle algorithm of Section 4.2 to compute sharp upper and lower bounds for the entries in this table corresponding to a number of different sets of fixed marginals. Here we describe one complex calculation for the set involving three fixed 15-way marginals obtained by collapsing n across the variables "managing money", "taking medicine" and "telephoning".

Of the $2^{16} = 65,536$ cells in the table, 62,384 contain zero entries. Since the target table is so sparse, releasing three marginals of dimension fifteen will lead to the exact disclosure of most of the cell entries. To be more exact, only 128 cells have the upper bounds strictly bigger than the lower bounds! The difference between the upper and lower bounds is equal to 1 for 96 cells, 2 for 16 cells, 6 for 8 cells, and 10 for 8 cells.

We take a closer look to the bounds associated with "small" counts of "1" or "2". A number of 1,729 cells contain a count of "1". From these, 1,698 cells have the upper bounds equal to the lower bounds. The difference between the bounds is 1 for 28 of the remaining counts of "1", is 2 for two other cells and is equal to 6 for only one entry. As for the 499 cells with a count of "2", the difference between the bounds is zero for 485 cells, is 1 for 10 cells and is 2 for 4 other cells. We need to emphasize that despite the tight bounds in this example, there may not be a disclosure concern for these data because they come from a sample and have been pooled across waves.

The generalized shuttle algorithm converged in approximately twenty iterations to the "correct" sharp bounds and it took less than six hours to complete on a single-processor machine at the Department of Statistics, Carnegie Mellon University. We re-checked these bounds by determining the feasible integer tables for which they are attained on the Terascale Computing System at the Pittsburgh Supercomputing Center. We used a parallel implementation of the shuttle algorithm and the computations took almost one hour to complete on fifty-six processors. We are currently exploring ways to speed up the calculations as well as approximations that will allow us to apply our results to larger tables.

5   Conclusions

In this paper we have explained how log-linear model statistical theory can help identify situations when explicit formulas exist for computing the best integer bounds on the entries of a cross-classification of arbitrary dimension given a set of marginal totals (the decomposable case). When such formulas do not exist, we illustrated how to derive similar formulas that help to reduce the computational effort (the reducible case). In addition, we explained how log-linear models provide the basis for correcting the shuttle algorithm originally proposed by Buzzigoli and Giusti, and transform it into a general procedure for computing sharp integer bounds given any set of marginals. The generalized shuttle algorithm described here simultaneously computes sharp integer bounds for all the cells by fully exploiting the structure of the bounds problem for multi-way contingency tables and, in addition, it can update the bounds, as more marginals are being released.

Acknowledgements

Preparation of this paper was supported in part by the National Science Foundation under Grant EIA-9876619 to the National Institute of Statistical

Sciences and by the National Institute of Aging under Grant 1090019 to Carnegie Mellon University. The computations were performed in part on the National Science Foundation Terascale Computing System at the Pittsburgh Supercomputing Center.

References

1. Bishop, Y.M.M., Fienberg, S.E., Holland, P.W. (1975). *Discrete Multivariate Analysis: Theory and Practice.* M.I.T. Press, Cambridge, MA
2. Bonferroni, C.E. (1936). *Teoria statistica delle classi e calcolo delle probabilita.* 8. Publicazioni del R. Instituto Superiore di Scienze Economiche e Commerciali di Firenze
3. Buzzigoli, L., Giusti, A. (1999). An Algorithm to Calculate the Lower and Upper Bounds of the Elements of an Array Given its Marginals. *Statistical Data Protection 1998 Proceedings*, 131–147, Eurostat, Luxembourg
4. Cox, L.H. (1999). Some Remarks on Research Directions in Statistical Data Protection. *Statistical Data Protection 1998 Proceedings*, 163–176, Eurostat, Luxembourg
5. Dobra, A. (2000). Computing Sharp Integer Bounds for Entries in Contingency Tables Given a Set of Fixed Marginals. Tech. Rep., Department of Statistics, Carnegie Mellon University
6. Dobra, A. (2000). Measuring the Disclosure Risk for Multi-way Tables with Fixed Marginals Corresponding to Decomposable Log-linear Models. Tech. Rep., Department of Statistics, Carnegie Mellon University
7. Dobra, A., Fienberg, S.E. (2000). Bounds for Cell Entries in Contingency Tables Given Marginal Totals and Decomposable Graphs. P. Natl. Acad. Sci. USA **97**, 11885–11892
8. Dobra, A., Fienberg, S.E. (2001). Bounds for Cell Entries in Contingency Tables Induced by Fixed Marginal Totals with Applications to Disclosure Limitation. Stat. J. United Nations ECE, **18**, 363–371
9. Duncan, G.T., Fienberg, S.E. (1999). Obtaining Information While Preserving Privacy: a Markov Perturbation Method for Tabular Data. *Statistical Data Protection 1998 Proceedings*, 351–362, Eurostat, Luxembourg
10. Edwards, D.E., Havranek, T. (1985). A Fast Procedure for Model Search in Multidimensional Contingency Tables. Biometrika **72**, 339–351
11. Erosheva, E. (2002). Grade of Membership and Latent Structure Models with Application to Disability Data. Department of Statistics, Carnegie Mellon University, Unpublished Ph.D. dissertation
12. Fienberg, S.E. (1999). Fréchet and Bonferroni Bounds for Multi-way Tables of Counts with Applications to Disclosure Limitation. *Statistical Data Protection 1998 Proceedings*, 115–129, Eurostat, Luxembourg
13. Fienberg, S.E., Makov, U.E. (1998). Confidentiality, Uniqueness and Disclosure Limitation for Categorical Data. J. Offic. Stat. **14**, 485–502
14. Fienberg, S.E., Makov, U.E., Meyer, M.M., Steele, R.J. (2001). Computing the Exact Distribution for a Multi-way Contingency Table Conditional on its Marginals Totals. *Data Analysis from Statistical Foundations: Papers in Honor of D. A. S. Fraser*, (Ed. Saleh, A.K.M.E.), 145–165, Nova Science Publishing, Huntington, NY

15. Fienberg, S.E., Makov, U.E., Steele, R.J. (1998). Disclosure Limitation Using Perturbation and Related Methods for Categorical Data. J. Offic. Stat. **14**, 485–502
16. Fréchet, M. (1940). *Les Probabilitiés, Associées a un Système d'Evénments Compatibles et Dépendants.* Vol. Premiere Partie. Hermann & Cie, Paris
17. Hoeffding, W. (1940). Scale-invariant correlation theory. Schriften des Mathematischen Instituts und des Instituts für Angewandte Mathematik der Universität Berlin **5**(3), 181–233
18. Joe, H. (1997). *Multivariate Models and Dependence Concepts.* Chapman & Hall, New York.
19. Keller-McNulty, S., Unger, E.A. (1998). A Database System Prototype for Remote Access to Information Based on Confidential Data. J. Offic. Stat. **14**, 347–360
20. Lauritzen, S.L. (1996). *Graphical Models.* Clarendon Press, Oxford
21. Leimer, H.G. (1993). Optimal Decomposition by Clique Separators. Discrete Math. **113**, 99–123
22. Roehrig, S.F., Padman, R., Duncan, G.T., Krishnan, R. (1999). Disclosure Detection in Multiple Linked Categorical Datafiles: a Unified Network Approach. *Statistical Data Protection 1998 Proceedings,* 149–162, Eurostat, Luxembourg
23. Samuels, S.M. (1998). A Bayesian, Species-sampling-inspired Approach to the Uniques Problem in Microdata Disclosure Risk Assessment. J. Offic. Stat. **14**, 373–384
24. Skinner, C.J., Holmes, D.J. (1998). Estimating the Re-identification Risk Per Record in Microdata. J. Offic. Stat. **14**, 361–372
25. Willenborg, L., de Waal, T. (1996). *Statistical Disclosure Control in Practice.* LNS **111**, Springer-Verlag, New York
26. Willenborg, L., de Waal, T. (2000). *Elements of Statistical Disclosure Control.* LNS **155**, Springer-Verlag, New York

Identification and Estimation with Incomplete Data

Joel L. Horowitz* and Charles F. Manski**

Department of Economics, Northwestern University, Evanston, Illinois, USA

Abstract. This paper is concerned with identification and estimation of econometric models when the sampling process produces missing observations. Missing observations occur frequently in applications due, for example, to non-response to questions on a survey or attrition from a panel. Missing observations usually cause population parameters of interest in applications to be unidentified except under untestable and often controversial assumptions. However, it is often possible to find identified, informative, bounds on these parameters that do not rely on untestable assumptions about the process through which data become missing. The bounds contain all logically possible values of the population parameters. Moreover, every parameter value within the bounds is consistent with some model of the process that generates missing observations. The bounds can be estimated consistently from data and often enable substantively important conclusions to be drawn without making untestable assumptions about missing observations. There are also situations in which the bounds are very wide. This is an indication that the data contain little information about the population parameters of interest and that substantive conclusions rely mainly on identifying assumptions that cannot be tested.

1 Introduction

Inference from incomplete data is a common problem in empirical research. For example, attrition from a panel and non-response to one or more questions on a survey are causes of missing observations and, therefore, incomplete data. Incomplete data also arise when responses specify only intervals that contain the variable of interest. For example, a survey may ask which of several intervals contains the respondent's income. In that case, the exact value of a respondent's income is missing.

Whatever the specific cause of incomplete data, the generic consequence is that the population parameters of interest in an application are not identified unless one makes untestable and frequently controversial assumptions about the distribution of missing data. For example, identification is possible if the missing and non-missing data have the same probability distribution

---

* The research of Joel L. Horowitz was supported in part by NSF grant SBR-9910925.

** The research of Charles F. Manski was supported in part by NSF grant SES-0001436.

or the same distribution conditional on some observed covariates. However, the hypothesis that missing and non-missing data have the same probability distribution cannot be tested and may not be plausible in a given application.

Although missing data make point identification problematic, it is often possible to identify informative bounds on population parameters without making untestable assumptions about missing data. The bounds on a parameter contain all logically possible values of that parameter (that is, all values that are consistent with the observed data and some process for generating the missing data), and they exhaust the information on the parameter that is available from the data. This paper describes completed and ongoing research on obtaining bounds for population parameters in the presence of missing data.

The analysis in this paper is deliberately conservative. We focus mainly on "worst case" scenarios in which the researcher has no prior information about the parameter of interest or the process that generates missing data. Our worst case approach contrasts with the "best case" approach that dominates the literature on inference from incomplete data. For example, it is a common practice to assume that data are missing completely at random (MCAR) and to perform analyses using only the non-missing data (e.g., survey responses in which all relevant questions were answered). Conventional methods for imputing missing data assume that missingness is random conditional on specified covariates. On occasion, a model of non-random missing data may be specified. Either way, the identification problem is solved, and efficiency of estimation becomes the central matter of concern to statisticians. We have emphasized in [2], [3], [5], and elsewhere that it is not sufficient for empirical researchers to know the inferences that can be made if specified assumptions hold. It is also important to be able to characterize the inferences that may be made without imposing these assumptions. An especially appealing feature of conservative analysis is that it enables establishment of a domain of consensus among researchers who may hold disparate beliefs about what assumptions are appropriate. A further important feature of our approach is that it provides an indication of the relative importance of the data and untestable assumptions in uniquely identifying the value of a parameter (point identification). If the identified bounds are narrow, then the data are highly informative about the parameter. However, if the bounds are wide, then the data contain little information about the parameter. Point identification must then rely heavily on untestable assumptions, and different assumptions can lead to very different identified values of the parameter.

Section 2 of this paper describes nonparametric identification of mean regressions when outcome or covariate data are missing. This section focuses on special cases that are important in applications and in which analytic formulae for bounds can be obtained. Section 3 describes ongoing research on the analysis of parameters that are general statistical functionals. In general, analytic expressions for bounds on such parameters are not available, and we describe numerical methods. Section 4 presents an empirical example, and Sect. 5 presents concluding comments.

To keep attention focused on the core problem of identification created by incomplete data, this paper does not dwell on the problem of estimation from finite samples. Most of the discussion supposes that the researcher knows the values of population features that are identified by the sampling process. In practice, bounds can usually be estimated consistently by replacing identified population features with sample analogs in analytic expressions for bounds or in numerical procedures for computing bounds.

2   Regression with Missing Outcome or Covariate Data

In this section, we consider a population whose members are characterized by the vector $W = (Y, X)$, where Y is a scalar outcome variable and X is a vector of covariates. The objective is to learn about the conditional expectation $E(Y|X \in A)$, where A is any measurable subset of the support of X. A random sample is drawn, but some data on (Y, X) are missing. We assume that $Y \in [0, 1]$. Boundedness of Y is necessary if worst-case inference of $E(Y|X \in A)$ is to yield informative conclusions in the presence of missing data. Given boundedness of Y, the restriction to the unit interval is a normalization that entails no loss of generality.

We begin by considering three extreme cases: only outcomes are missing (Section 2.1), outcomes and covariates are jointly missing (Section 2.2), and only covariates are missing (Section 2.3). Section 2.4 discusses more general patterns of missing data. An empirical illustration is given in Section 4.

2.1   Missing Outcome Data

Manski [4] analyzed the case in which X is always observed but data on Y may be missing. Let $Z = 1$ if (Y, X) is observed, and $Z = 0$ if only X is observed. Then

$$\mathbf{E}(Y|X \in A) = \mathbf{E}(Y|X \in A, Z = 1)\mathbf{P}(Z = 1|X \in A) \tag{1}$$
$$+ \mathbf{E}(Y|X \in A, Z = 0)\mathbf{P}(Z = 0|X \in A)$$

The quantities $\mathbf{E}(Y|X \in A, Z = 1)$, $\mathbf{P}(Z = 1|X \in A)$, and $\mathbf{P}(Z = 0|X \in A)$ are identified by the sampling process, but $\mathbf{E}(Y|X \in A, Z = 0)$ is not. The last quantity can have any value in $[0, 1]$. Therefore, we obtain the identification region

$$\mathbf{E}(Y|X \in A, Z = 1)\mathbf{P}(Z = 1|X \in A) \tag{2}$$
$$\leq \mathbf{E}(Y|X \in A)$$
$$\leq \mathbf{E}(Y|X \in A)\mathbf{P}(Z = 1|X \in A) + \mathbf{P}(Z = 0|X \in A).$$

These bounds are sharp. That is, they contain every value of $E(Y|X \in A)$ that is consistent with the identified quantities ($E(Y|X \in A, Z = 1)$, $P(Z = 1|X \in A)$, and $P(Z = 0|X \in A)$) and some value of the unidentified quantity $E(Y|X \in A, Z = 0)$. Moreover, every value of $E(Y|X \in A)$ that is within the bounds is consistent with the values of the identified quantities for some value of $E(Y|X \in A, Z = 0)$. Thus, the bounds exhaust the information about $E(Y|X \in A)$ that is available from the data. The width of the interval containing $E(Y|X \in A)$ increases from 0 to 1 as the probability of a non-missing observation, $P(Z = 1|X \in A)$ decreases from 1 to 0.

2.2   Jointly Missing Outcome and Covariate Data

Horowitz and Manski [2] analyzed the case in which some realizations of (Y, X) are entirely missing and the rest are fully observed. Let $Z = 1$ if (Y, X) is observed and $Z = 0$ otherwise. It follows from Bayes' theorem that

$$E(Y|X \in A) \tag{3}$$
$$= E(Y|X \in A, Z = 1)\frac{\pi(A, 1)P(Z = 1)}{\pi(A, 1)P(Z = 1) + \pi(A, 0)P(Z = 0)}$$
$$+ E(Y|X \in A, Z = 0)\frac{\pi(A, 0)P(Z = 0)}{\pi(A, 1)P(Z = 1) + \pi(A, 0)P(Z = 0)}.$$

where $\pi(A, j) = P(X \in A|Z = j)$. The quantities $E(Y|X \in A, Z = 1)$, $\pi(A, 1)$, and $P(Z = j)$ are identified by the sampling process, but $E(Y|X \in A, Z = 0)$ and $\pi(A, 0)$ are not. The identification bounds for $E(Y|X \in A)$ are obtained by maximizing and minimizing the right-hand side of (3) over all values of $E(Y|X \in A, Z = 0) \in [0, 1]$ and $\pi(A, 0) \in [0, 1]$. The result is

$$E(Y|X \in A, Z = 1)P_e(Z = 1|X \in A) \tag{4}$$
$$\leq E(Y|X \in A)$$
$$\leq E(Y|X \in A, Z = 1)P_e(Z = 1|X \in A) + P_e(Z = 1|X \in A),$$

where

$$P_e(Z = 1|X \in A) = \frac{P(A, 1)P(Z = 1)}{\pi(A, 1)\pi(Z = 1) + P(Z = 0)}$$

is the *effective response probability,* and $P_e(Z = 0|X \in A) = 1 - P_e(Z = 1|X \in A)$.

The bounds (4) have the same form as (2), except the identified effective response probability $P_e(Z = 1|X \in A)$ replaces the unidentified probability $P(Z = 1|X \in A)$. The width of the identification region increases from 0 to 1 as either $\pi(A, 1)$ or $P(Z = 1)$ decreases from 1 to 0.

2.3 Missing Covariate Data

Horowitz and Manski [2] also analyzed the case in which Y is always observed but data on X may be missing. Now $Z = 1$ if (Y, X) is observed, and $Z = 0$ if only Y is observed. This case is more complex than those discussed so far.

To derive the identification region, reconsider the analysis of jointly missing outcome and covariate data in Sect.2.2. There the available data constrained the right-hand side of (3) by identifying $\mathbf{E}(Y|X \in A, Z = 1)$, $\pi(A, 1)$, and $\mathbf{P}(Z = j)$. If only covariate data are missing, the data also identify $\mathbf{P}(Y|Z = 0)$. Knowledge of this probability jointly constrains $\mathbf{E}(Y|X \in A, Z = 0)$ and $\pi(A, 0)$ through the equation

$$\mathbf{P}(Y|Z = 0) = \mathbf{P}(Y|X \in A, Z = 0)\pi(A, 0) + \mathbf{P}(Y|X \in \bar{A}, Z = 0)\pi(\bar{A}, 0), \quad (5)$$

where $\bar{A}$ denotes the complement of A. To determine the implications of (5), let $p \in [0, 1]$, and suppose that $\pi(A, 0) = p$. Let Ψ denote the set of all possible distributions of Y. Then the values of $\mathbf{P}(Y|X \in A, Z = 0)$ that are consistent with (5) are

$$\Psi(p) = \Psi \cap \{[\mathbf{P}(Y|Z = 0) - (1 - p)\psi]/p : \psi \in \Psi\}. \quad (6)$$

The implied set of feasible values for $\mathbf{E}(Y|X \in A, Z = 0)$ is

$$\mathbf{E}(Y|X \in A, Z = 0) \in [g_0(p), g_1(p)], \quad (7)$$

where

$$g_0(p) = \inf \left[\int y d\psi : \psi \in \Psi(p) \right]$$

and

$$g_1(p) = \sup \left[\int y d\psi : \psi \in \Psi(p) \right].$$

It can be shown that $g_0(p)$ and $g_1(p)$ are the means of two truncated versions of $\mathbf{P}(Y|Z = 0)$, specifically the distributions formed from the left and right tails containing mass p (see [1]). Combining (3) and (7) yields

$$\mathbf{E}(Y|X \in A, Z = 1)\frac{p\mathbf{P}(Z = 1)}{p\mathbf{P}(Z = 1) + (1 - p)\mathbf{P}(Z = 0)} \quad (8)$$

$$+ g_0(p)\frac{p\mathbf{P}(Z = 1)}{p\mathbf{P}(Z = 1) + (1 - p)\mathbf{P}(Z = 0)}$$

$$\leq \mathbf{E}(Y|X \in A)$$

$$\leq \mathbf{E}(Y|X \in A, Z = 1)\frac{p\mathbf{P}(Z = 1)}{p\mathbf{P}(Z = 1) + (1 - p)\mathbf{P}(Z = 0)}$$

$$+ g_1(p)\frac{p\mathbf{P}(Z = 1)}{p\mathbf{P}(Z = 1) + (1 - p)\mathbf{P}(Z = 0)}.$$

If it were known that $\pi(A,0) = p$, then the right-hand side of (8) would give the identification bounds for $\mathbf{E}(Y|X \in A)$. However, the sampling process places no restrictions on $\pi(A,0)$. Therefore, the identification bounds for $\mathbf{E}(Y|X \in A)$ are formed by the union over $p \in [0,1]$ of the intervals on the right-hand side of (8).

In general, the resulting region does not have a simple analytic form comparable to those given in Sects. 2.1 and 2.2. However, one special case yields an exceedingly simple and surprising result. Suppose that $\mathbf{P}(Y|Z = 0)$ is degenerate with all mass at $\mathbf{E}(Y|X \in A, Z = 1)$. Then $g_0(p) = g_1(p) = \mathbf{E}(Y|X \in A, Z = 1)$ for all $p \in [0,1]$. Therefore, (8) reduces to $\mathbf{E}(Y|X \in A) = \mathbf{E}(Y|X \in A, Z = 1)$. Thus, $\mathbf{E}(Y|X \in A)$ is identified even if X is never observed.

2.4   General Patterns of Missing Data

Analytic characterization of identification bounds for $\mathbf{E}(Y|X \in A)$ is possible only in special cases when some sample realizations may have missing outcome data, others may have missing covariate data, and still others have jointly missing outcomes and covariates. The general problem is discussed in Sect. 3. Here, we treat a special case where analytic results are available.

Let Y be a binary outcome variable so that $\mathbf{E}(Y|X \in A) = \mathbf{P}(Y|X \in A)$. Suppose that covariate data are either entirely observed or entirely missing. Thus, if X is a vector, then either all of its components are observed or all are missing. Let Z_y and Z_x be indicators of missing data. Y is observed if $Z_y = 1$ and missing if $Z_y = 0$. X is observed if $Z_x = 1$ and missing if $Z_x = 0$. Define $\Pi(A) = \mathbf{E}(Y|X \in A)$. For $j, k = 0, 1$, define $E_{jk} = \mathbf{E}(Y|X \in A, Z_x = j, Z_y = k)$, $A_{jk} = \mathbf{E}(Y|Z_x = j, Z_y = k)$, $G_{jk} = \mathbf{P}(Z_x = j, Z_y = k|X \in A)$, $Q_{jk} = \mathbf{P}(X \in A|Z_x = j, Z_y = k)$, and $p_{jk} = \mathbf{P}(Z_x = j, Z_y = k)$. Then

$$\Pi(A) = \sum_j \sum_k E_{jk} G_{jk}. \tag{9}$$

Application of Bayes' theorem to G_{jk} in (9) gives

$$\Pi(A) = \sum_j \sum_k E_{jk} Q_{jk} p_{jk} \Big/ \sum_j \sum_k Q_{jk} p_{jk}. \tag{10}$$

For $j, k \in \{0, 1\}$, the sampling process identifies E_{11}, Q_{1k}, p_{jk}, and A_{01}. It does not identify Q_{0k}, E_{j0}, or E_{01}. These quantities can have any values in $[0,1]$. However,

$$A_{01} = E_{01} Q_{01} + B(1 - Q_{01}), \tag{11}$$

where $B = \mathbf{E}(Y|X \neq x, Z_x = 0, Z_y = 1)$. B is not identified and can have any value in $[0,1]$. Therefore, it follows from (11) that

$$\max\left[0, \frac{A_{01} - (1 - Q_{01})}{Q_{01}}\right] \leq E_{01} \leq \min\left(1, \frac{A_{01}}{Q_{01}}\right). \tag{12}$$

These bounds are sharp (see [1], Corollary 1.2). Therefore, sharp bounds on $\Pi(A)$ can be obtained by minimizing and maximizing the right-hand side of (10) with respect to the unidentified quantities subject to $0 \leq Q_{0k} \leq 1$, $0 \leq E_{j0} \leq 1$ $(j, k = 0, 1)$, and (12). To state the result, define

$$D = \sum_{k=0}^{1} Q_{1k}p_{1k} + p_{00} + (1 - A_{01})p_{01},$$

$$R = \sum_{k=0}^{1} Q_{1k}p_{1k} + p_{00} + A_{01}p_{01},$$

$S = E_{11}Q_{11}p_{11} + Q_{10}p_{10} + p_{00} + A_{01}p_{01}$, $L = E_{11}Q_{11}p_{11}/D$, and $U = S/R$. Then as is shown in [3], sharp bounds on $\Pi(x)$ are

$$L \leq \Pi(A) \leq U. \tag{13}$$

A necessary and sufficient condition for the bounds in (13) to be informative (that is, to satisfy $U - L < 1$) is $Q_{11} > 0$ and $p_{11} > 0$. In other words, the bounds are informative if the probability of a complete observation with $X \in A$ exceeds zero.

Horowitz and Manski [3] also find sharp identification bounds for the contrast $\Pi(B) - \Pi(A) \equiv E(Y|X \in B) - E(Y|X \in A)$, where B and A are any two disjoint subsets of the support of X. The analysis is subtle because a missing covariate realization cannot be in A and B simultaneously. Therefore, the identification region for $\Pi(B) - \Pi(A)$ is a proper subset of the region formed by considering all logically possible values of $\Pi(B)$ and $\Pi(A)$. To state the result, let $E_{B_{jk}}$ and $Q_{B_{jk}}$ be the quantities obtained from E_{jk} and Q_{jk} by replacing A with B. Define $a = E_{B11}Q_{B11}p_{11} + Q_{B10}p_{10} + A_{01}p_{01}$, $b = Q_{B11}p_{11} + Q_{B10}p_{10} + A_{01}p_{01}$, $d = E_{11}Q_{11}p_{11}$, $f = Q_{11}p_{11} + Q_{10}p_{10} + (1 - A_{10})p_{01}$,

$$G(z) = \frac{p_{00}(b - a)}{(b + p_{00}z)^2} - \frac{p_{00}d}{[f + p_{00}(1 - z)]^2},$$

and

$$z^* = \begin{cases} 1 & \text{if } G(0) > 0 \text{ and } G(1) > 0, \\ 0 & \text{if } G(0) < 0 \text{ and } G(1) < 0, \\ \text{the solution in } [0, 1] \text{ to } G(z) = 0, \text{ otherwise.} \end{cases}$$

Then sharp bounds on $\Pi(B) - \Pi(A)$ are

$$L_{BA} \leq \Pi(B) - \Pi(A) \leq U_{BA}, \tag{14}$$

where $U_{BA} = \dfrac{a + p_{00}z^*}{b + p_{00}z^*} - \dfrac{d}{f + p_{00}(1 - z^*)}$, $L_{BA} = -U_{AB}$, and U_{AB} is obtained from U_{BA} by exchanging B and A (see [3]).

Tighter bounds can be obtained if X is missing completely at random (MCAR). Although the hypothesis that X is MCAR cannot be tested, there are applications in which enough is known about the causes of missing observations of X to make MCAR a reasonable assumption. See [3] for an example. Formally, X is MCAR if $\mathbf{P}(Z_x = j|Y = \ell, X \in A, Z_y = k) = \mathbf{P}(Z = j)$ for all subsets A of the support of X and all $j, k, \ell = 0, 1$. Define $H = \mathbf{P}(Z_y = 1|X \in A, Z_x = 1)$, $L_m = E_{11}H$, and $U_m = 1 - H + L_m$. Horowitz and Manski ([3]) show that if X is MCAR, then sharp bounds on $\Pi(A)$ are

$$L_m \leq \Pi(A) \leq U_m. \tag{15}$$

The MCAR condition can also be used to obtain bounds on $\Pi(B) - \Pi(A)$ that are tighter than those given in the previous paragraph. To state these, let H_B be the quantity that is obtained from H by replacing A with B. Define $U_{BAm} = 1 - (1 - E_{B11})H_B - E_{11}H$ and $L_{ABm} = -U_{ABm}$, where U_{ABm} is obtained from U_{BAm} by exchanging A and B. Horowitz and Manski ([3]) show that if X is MCAR, then sharp bounds on $\Pi(B) - \Pi(A)$ are

$$L_{BAm} \leq \Pi(B) - \Pi(A) \leq U_{BAm}. \tag{16}$$

It is not difficult to show that the bounds in (15) and (16) are informative even when B and A are sets of measure zero. In contrast, the bounds in (13) and (14), which do not assume that X is MCAR, are uninformative when B and A are sets of measure zero. That is, the lower and upper bounds in (13) are 0 and 1, and the lower and upper bounds in (14) are -1 and 1 when B and A are sets of measure zero.

3   Identification Bounds on General Statistical Functionals

Most population parameters of interest in applications can be expressed as statistical functionals. Specifically, let F be the cumulative distribution function (CDF) of a random variable X in the sampled population. Then a parameter θ typically can be written in the form $\theta = G(F)$ for some known functional G. The problem is to infer the scalar quantity $h(\theta)$, where h is a known function. For example, if θ is a vector, then $h(\theta)$ might be one of its components.

This framework encompasses a large class of estimation problems. For example, unconditional and conditional means and medians can be written as statistical functionals. The same is true for parameters that are identified by the solutions to extremum problems. Other familiar examples are the best linear predictor (BLP) of the conditional mean of Y and the "slope" coefficients of a binary logit or probit model. To illustrate, the BLP of $\mathbf{E}(Y|X = x)$ is $x\theta$, where

$$\theta = (\mathbf{E}X'X)^{-1}\mathbf{E}X'Y = \left(\int x'x dF_{yx}\right)^{-1} \Big/ \left(\int x'y dF_{yx}\right), \tag{17}$$

where X is a row vector of explanatory variables and F_{yx} is the joint CDF of (Y, X). In a binary logit model, $P(Y = 1|X = x) = \exp(x\theta)/[1 + \exp(x\theta)]$, and it is not difficult to show that

$$\theta = (\mathbf{E}X'X)^{-1}\{\mathbf{E}X' \log[P(Y = 1|X)/P(Y = 0|X)]\}$$
$$= \left(\int x'x dF_{yx}\right)^{-1} \bigg/ \int \{x' \log[P(Y = 1|X = x)/P(Y = 0|X = x)]\} dF_{yx}.$$

When estimating statistical functionals, empirical researchers routinely report estimates based only on sample realizations that are completely observed. This practice is justified if the same population probability distribution generates the realizations that are completely and incompletely observed. Otherwise, it is usually not justified and can produce seriously misleading results. To illustrate, consider the BLP of $\mathbf{E}(Y|X = x)$. Let $Z = 1$ if (Y, X) is completely observed and $Z = 0$ otherwise. Let $F_{yx|z}$ denote the CDF of (Y, X) conditional on $Z = 1$. Then standard practice is to estimate $x\theta_c$ instead of $x\theta$, where

$$\theta_c = \left(\int x'x dF_{yx|z}\right)^{-1} \bigg/ \left(\int x'y dF_{yx|z}\right).$$

Except in special cases, $\theta_c \neq \theta$ unless $F_{yx} = F_{yx|z}$ almost surely, in which case observations are missing completely at random. It is important to know what can be learned about a parameter of interest when the researcher has no prior information about the distribution of the missing data or the process that causes data to be missing. This motivates the research that is described in the remainder of this section.

Let $V = (Y, X)$ be a random vector. Typically, Y is a scalar dependent variable and $X \in \mathbb{R}^d$ is a vector of explanatory variables, but this distinction is not necessary for the general formulation that is presented in this section. We assume that θ is a continuous functional of the CDF of V and that V is a discrete random variable with support $\{\nu_i : i = 1, \ldots, I\}$. The assumption that V is discrete entails no significant loss of generality, because the CDF of a continuously distributed random variable can be approximated with arbitrary accuracy by a discrete CDF. Let Z $(1 \leq Z \leq z_{\max})$ be an integer-valued random variable that indicates the state of missingness of V. Define $Z = 1$ if all components of V are observed and $Z = z_{\max}$ if none are observed. Intermediate values of Z indicate combinations of components of V that are observed and missing. For example $Z = 2$ might indicate that all components of V but the last are observed. Also define $\pi_z = P(Z = z)$ and $p_{zi} = P(V = \nu_i|Z = z)$. The sampling process identifies π_z for all $z \in (1, \ldots, z_{\max})$ and p_{1i} for all $i = 1, \ldots, I$. The remaining p_{zi}'s are not identified. However, there are restrictions on the values of the unidentified p_{zi}'s.

To obtain these, define S_z to be the support of the components of V that are observed (non-missing) when $Z = z$. Suppose that there are K_z distinct points in S_z. Let q_{zk_z} denote the marginal probability of point $k_z \in S_z$

conditional on $Z = z$. These marginal probabilities are identified by the sampling process. When $Z = z$, write $V_i \in k_z$ if the non-missing components of V correspond to the point $k_z \in S_z$. Then the following relations hold:

$$\sum_{i:\nu_i \in k_z} p_{zi} = q_{zk_z} \qquad (z = 2, \ldots, z_{\max} - 1; k_z = 1, \ldots, K_z) \qquad (18a)$$

$$\sum_{i=1}^{I} p_{zi} = 1 \qquad (z = 2, \ldots, z_{\max}) \qquad (18b)$$

$$p_{zi} \geq 0 \qquad (z = 2, \ldots, z_{\max}; i = 1, \ldots, I) \qquad (18c)$$

In addition, the probability mass function corresponding to F can be written in the form

$$\mathbf{P}(V = \nu_i) = \sum_{z=1}^{z_{\max}} \pi_z p_{zi}.$$

Since the probability mass function determines F uniquely, θ can be written in the form

$$\theta = g\left(\sum_{z=1}^{z_{\max}} \pi_z p_{z1}, \sum_{z=1}^{z_{\max}} \pi_z p_{z2}, \ldots, \sum_{z=1}^{z_{\max}} \pi_z p_{zI}\right).$$

Therefore, $h(\theta)$ has the form

$$h(\theta) = h\left[g\left(\sum_{z=1}^{z_{\max}} \pi_z p_{z1}, \sum_{z=1}^{z_{\max}} \pi_z p_{z2}, \ldots, \sum_{z=1}^{z_{\max}} \pi_z p_{zI}\right)\right].$$

The identification problem is now clear: $h(\theta)$ depends on probabilities $p_{zi}(z \geq 2)$ that are not identified by the sampling process. The identified sharp bounds on $h(\theta)$ are the maximum and minimum values of $h(\theta)$ that are consistent with the constraints (18a) – (18c). Thus, the bounds are the optimal solutions to

(NLP)   minimize (maximize):
$$\underset{p_{zi}:z \geq 2; i=1,\ldots,I}{}$$

$$h(\theta) = h\left[g\left(\sum_{z=1}^{z_{\max}} \pi_z p_{z1}, \sum_{z=1}^{z_{\max}} \pi_z p_{z2}, \ldots, \sum_{z=1}^{z_{\max}} \pi_z p_{zI}\right)\right]$$

subject to: $$\sum_{i:\nu_i \in k_z} p_{zi} = q_{zk_z} \qquad (z = 2, \ldots, z_{\max} - 1; k_z = 1, \ldots, K_z)$$

$$\sum_{i=1}^{I} p_{zi} = 1 \qquad (z = 2, \ldots, z_{\max})$$

$$p_{zi} \geq 0 \qquad (z = 2, \ldots, z_{\max}; i = 1, \ldots, I)$$

(NLP) is a mathematical programming problem with linear constraints and a nonlinear objective function. It can be solved analytically in special cases such as those described in Sect. 2. In general, however, analytic solution is not possible and numerical methods must be used. The difficulty of solving (NLP) numerically depends on the details of the objective function. This is an area of ongoing research. Section 4 presents an empirical example based on solving (NLP).

Before presenting the example, we note that any information about the distribution of missing data or the process through which data become missing can be incorporated into (NLP) by adding constraints. For example, suppose that V can be partitioned (V_a, V_b) and that V_b is MCAR. Let $Z_a = 1$ if V_a is observed and $Z_a = 0$ otherwise. Let $Z_b = 1$ if V_b is observed and $Z_b = 0$ otherwise. Define $Z = (Z_a, Z_b)$. The assumption that V_b is MCAR implies that

$$\mathbf{P}(V = \nu_i | Z_a = z_a, Z_b = z_b) = \mathbf{P}(V = \nu_i | Z_a = z_a)$$

for $z_a, z_b = 0$ or 1. This is equivalent to

(M) $$p_{zi} = \sum_{\zeta \in \{z : z_a = A\}} \pi_\zeta p_{\zeta i} \quad (i = 1, ..., I; A = 0, 1).$$

Therefore, the assumption that V_b is MCAR can be incorporated in (NLP) by adding constraint (M).

4 An Empirical Example

This section presents an empirical example that illustrates the ideas developed in Sects. 2 – 3. Undercover agents of the U.S. Drug Enforcement Administration and the Metropolitan Police of the District of Columbia buy cocaine to use as evidence in criminal investigations. This section presents an example in which the methods described in Sect. 3 are used to compute bounds on the BLP of the mean of the logarithm of the cost of cocaine conditional on the logarithm of the quantity purchased. Specifically, bounds are found on the parameter θ_1 in the model

$$\log C = \theta_0 + \theta_1 \log Q + U,$$

where C is cost in dollars, Q is quantity in grams, and U is an unobserved random variable. The population values of θ_0 and θ_1 minimize $\mathbf{E}(\log C - \theta_0 - \theta_1 \log Q)^2$.

The data are records of purchases of cocaine powder in 1986. There are 409 records but only 321 are complete. The pattern of missingness is shown in Tab. 1. Table 2 shows bounds on θ_1 that are computed (a) without making any assumptions about missing observations, (b) under the assumption that $\log Q$ is MCAR, and (c) under the assumption that $\log C$ and $\log Q$ are both MCAR. We have no reason for believing that either $\log C$ or $\log Q$ is MCAR, but we present bounds under these assumptions to illustrate their effects on the estimates. As expected, the bounds are widest when no assumptions are made about the missing data and narrower when $\log Q$ is assumed to be MCAR. When $\log C$ and $\log Q$ are both assumed to be MCAR, θ_1 is point identified. Even under no assumptions about missingness, it is clear that $\theta_1 < 1$. Thus, the cost per unit of cocaine decreases as the quantity purchased increases. In other words, there is quantity discounting in the market for cocaine. This is a substantively important finding that is consistent with the results of many other studies of markets for illegal drugs. The result shows that the procedures of Sect. 3 can be used to obtain substantively important results without making untestable assumptions about missing data.

5   Conclusions

Missing or incomplete data cause population quantities of interest to be unidentified unless untestable assumptions are made about the probability distribution of the missing data. This paper has argued that sharp, informative bounds on population parameters are often available without making untestable assumptions about missingness. The bounds exhaust the information that is available from the data. In some cases, they can be calculated analytically. More generally, the bounds are solutions to nonlinear mathematical programming problems and must be computed numerically. This paper has shown that the computations are tractable in some important leading cases, but further research is necessary to develop a complete understanding of the computational issues that are involved in solving the general mathematical programming problem. An example based on real data has illustrated the ability of the bounding procedure to provide substantively useful results without making untestable assumptions about missing data.

Table 1. Pattern of missingness in the cocaine data

	Complete records	Missing only cost	Missing only quantity	Missing cost and quantity
Number	321	35	43	10
Percent	78	9	11	2

Table 2. Bounds on θ_1 in cocaine example

Assumption about missingness	Lower Bound	Upper Bound
None	0.03	0.51
Quantity is MCAR	0.20	0.47
Cost and *Quantity* are MCAR	0.39	0.39

References

1. Horowitz, J.L., Manski, C.F. (1995). Identification and Robustness with Contaminated and Corrupted Data, Econometrica, **63**, 281–302
2. Horowitz, J.L., Manski, C.F. (1998). Censoring of Outcomes and Regressors Due to Survey Nonresponse: Identification and Estimation Using Weights and Imputations, J. Econometrics, **84**, 37–58
3. Horowitz, J.L., Manski, C.F. (2000). Nonparametric Analysis of Randomized Experiments with Missing Covariate and Outcome Data, J. Am. Stat. Assoc., **95**, 77–84
4. Manski, C.F. (1989). Anatomy of the Selection Problem, J. Hum. Resour., **24**, 343–360
5. Manski, C.F. (1995). *Identification Problems in the Social Sciences*. Harvard University Press, Cambridge

Computational Information Retrieval

Jacob Kogan*

Department of Mathematics and Statistics, University of Maryland, Baltimore
County, Baltimore, USA

Abstract. The main goal of this note is to introduce the notion of collection
dependent "same context words". Two (or more) words are the "same context
words" if they occur in the same (or similar) context across a given text collection.
Each word w in the collection is associated with a profile $\mathcal{P}(w)$. The profile $\mathcal{P}(w)$
is the set of words occurring in sentences that contain w. We introduce a distance
function in the set profiles, and use it to cluster words. Words contained in the
same cluster are "same context words". We select "same context words" for several
text collections, and briefly discuss further possible applications of the introduced
concepts to a number of information retrieval related problems.

1 Introduction

A common form of text processing in many information retrieval systems
is based on analysis of word occurrences across a document collection. The
number of words used by the system defines the dimension of a vector space
in which the analysis is carried out. Reduction of the dimension may lead to
significant savings of computer resources and processing time. At the same
time the savings may dramatically degrade the quality of retrieval.

Stemming is on of the best known general methods to reduce the number
of collection unique words. While a number of efficient stemming algorithms
are already available they are not specifically designed to conflate words having similar meanings. This, in turn, may lead to occasional retrieval failures
(see e.g. [20]). Latent Semantic Indexing (see e.g. [5], [9] [13]) is another
noticeable dimension reduction technique based on linear algebra tools. We
believe that the "same context words" analysis presented in the paper (along
with stop list removal and stemming) should be used for the initial vector
space model construction. This construction should precede possible following transformations of the vector space. For this reason we do not discuss
LSI in the paper.

Rather than consider general–purpose language tools the paper introduces
corpus based measure of similarity between words (for detailed motivation of
text dependent approaches in information retrieval we refer to e.g. [20]). Our
departure point is the definition (attributed to Leibniz): two expressions are
synonymous if the substitution of one for the other never changes the truth
value of a sentence in which the substitution is made. Similar recent claims

* This research was partially supported by CyberTavern.TV LLC

"you can begin to know the meaning of a word (or term) by the company it keeps" and "words or terms that occur in 'the same context' are 'equivalent'" (see [11]), and "the assumption is that words with similar meanings will occur with similar neighbors if enough text material is available" (see [18]) provide an additional motivation for this line of research.

We argue that in large text collections words with similar meanings could be found in similar contexts. The argument, if true, would imply that words that occur in sentences together with a word w may provide valuable information concerning w. We define a profile $\mathcal{P}(w)$ of a word w as the set of words that occur in sentences containing w (for details see Section 2). We introduce a distance function in the set of profiles so that the similarity between two words w_1 and w_2 is measured by the distance between the corresponding profiles $\mathcal{P}(w_1)$, and $\mathcal{P}(w_2)$. We use the similarity measure between words to partition corpus words into clusters. Words in the same cluster are "same contexts words."

"Same context words" w' and w'' do not have to occur in the same sentence, our experiments show that, for example, the pairs {wonder, miracle}, and {fight, battle} are same contexts words for "A Connecticut Yankee in King Arthur's Court" by Mark Twain. Same contexts words should not be synonyms. The pairs {mean, kind}, or {shield, horse} are also "same contexts words" for the same text. "Same context words", in general, do not constitute association rules (see [1]).

The paper is organized as follows. Definitions and methodology are presented in Section 2. Section 3 contains summary of experiments with four different text collections. A partial list of potential applications is presented in Section 4. Brief conclusions are given in Section 5.

2   Profiles

Following Kowalski [14] we apply the following pre–processing operations to create a searchable data structure from a text collection **T**:

1. stop list words removal (the stop list is available from
   ftp://ftp.cs.cornell.edu/pub/smart/english.stop),
2. punctuation removal,
3. translation of upper case characters into lower case.

Stemming algorithms strip word's ending and often conflate words with common roots. In this paper we shall call an output of a stemming algorithm a term. For example, an application of Porter stemming algorithm (to be discussed in Section 4) generates term "studi" from each one of the words "study", "studying", and "studied". To avoid editing of clustering results presented in Section 3 we do not apply stemming to documents discussed in this section.

Next we construct a sorted list of unique words $\mathbf{w} = \{w_1, \ldots, w_n\}$ that appear in the text collection $\mathbf{T}$. For each word w on the list $\mathbf{w}$ we denote the set of sentences in $\mathbf{T}$ containing w by $\mathbf{s}(w)$. For each word $w \in \mathbf{w}$ we define profile $\mathcal{P}(w)$ as follows:

Definition 1. The profile $\mathcal{P}(w)$ of the word w is an alphabetically sorted list of words from the list $\mathbf{w}$ that occur in sentences together with the word w.

Information concerning the word's w "company" is contained in the profile $\mathcal{P}(w)$. There is a number of ways to associate a vector $\mathbf{p}(w) = \{p_1, \ldots, p_n\}^T$ with the profile $\mathcal{P}(w)$. For example, one can define the i−th coordinate p_i of the vector $\mathbf{p}(w)$ as the number of sentences in the document collection that contain both words w and w_i. The definition when applied, for example, to Reuters business news collection (to be discussed in details in Section 3) identifies the following words as "same context words":

$$\{\text{white, house}\}, \{\text{margaret, thatcher}\}, \{\text{president, reagan}\}.$$

In order to avoid this type of "same context words" we exclude the "contribution" of the word w into the vector $\mathbf{p}(w)$ and define the profile vector $\mathbf{p}(w)$ as follows:

Definition 2. The vector $\mathbf{p}(w)$ of the profile $\mathcal{P}(w)$ is the vector in $\mathbf{R}^n$

$$\mathbf{p}(w) = \{p_1, \ldots, p_n\}^T,$$

where

$$p_i = \begin{cases} \text{the number of sentences in } \mathbf{s}(w) \text{ containing } w_i \text{ if } w_i \neq w \\ 0 \qquad\qquad\qquad\qquad\qquad\qquad\qquad\qquad\ \text{if } w_i = w. \end{cases}$$

The definition of the vector $\mathbf{p}(w)$ is motivated by the desire to distinguish between two different words w' and w'' that often co-occur in sentences across a document collection (like, for example, "white" and "house" in the above discussion). While the profiles $\mathcal{P}(w')$ and $\mathcal{P}(w'')$ may be identical, the corresponding profile vectors $\mathbf{p}(w')$ and $\mathbf{p}(w'')$ defined above are different.

To illustrate the definitions we introduce an example.

Example 1. Consider the following text $\mathbf{T}$ (see [4], p.1):
"We expect a lot from our search engines. We ask them vague questions about topics that we're unfamiliar with ourselves and in turn anticipate a concise, organize response. We type in principal when we meant principle."

1. An application of the stop list words removal transforms the text into:
    "expect lot search engines. ask vague questions topics that unfamiliar turn anticipate concise, organize response. type principal meant principle."

2. Punctuation removal generates:
   "expect lot search engines ask vague questions topics that unfamiliar turn anticipate concise organize response type principal meant principle"
3. Finally, translation of upper case characters into lower case and sorting produces the sorted list of of unique words $\mathbf{w} = \{w_1, \ldots, w_n\}$ for the text:
   {anticipate, concise, engines, expect, lot, meant, organize, principal, principle, questions, response, search, topics, turn, type, unfamiliar, vague}.

Since the list $\mathbf{w}$ contains 17 words, for each $w \in \mathbf{w}$ the profile $\mathbf{p}(w) \in \mathbf{R}^{17}$. For the word $w =$ "principle" the set of sentences $\mathbf{s}($ "principle"$)$ contains one sentence

$$\mathbf{s}(\text{"principle"}) = \{\text{We type in principal when we meant principle.}\}.$$

The profile of w is

$$\mathcal{P}(\text{"principle"}) = \{\text{meant, principal, principle, type}\},$$

the vector $\mathbf{p}($ "principle"$)$ associated with the profile is

$$\mathbf{p}(\text{"principle"}) = (0,0,0,0,0,1,0,1,0,0,0,0,0,0,1,0,0)^T.$$

The coordinates $p_6 = p_8 = p_{15} = 1$ and correspond to $w_6 = $ "meant", $w_8 = $ "principal", $w_{15} = $ "type". The coordinate p_9 corresponding to $w_9 = $ "principle" is zero.

Statistical approach to the data forces us to disregard words with small profiles. Specifically, we introduce a parameter MinNumSent and select a "cut"-a set of words w whose set $\mathbf{s}(w)$ is at least as large as the parameter:

$$\text{cut} = \text{cut}\,(\text{MinNumSent}) = \{w \,:\, w \in \mathbf{w},\ |\mathbf{s}(w)| \geq \text{MinNumSent}\}.$$

Most of the experiments described in the paper are conducted with cuts.

In this paper we use a modified k–means algorithm with the cosine similarity measure (for detailed description of the algorithm see [16]). Hence, instead of the profile vectors $\mathbf{p}(w)$, we shall use normalized profile vectors $\mathbf{p_n}(w)$. That is $\mathbf{p_n}(w) = \dfrac{\mathbf{p}(w)}{|\mathbf{p}(w)|_2}$, where for two vectors $\mathbf{p}, \mathbf{q} \in \mathbf{R}^n$, we denote the dot product $\sum_{i=1}^{n} p_i q_i$ by $\mathbf{p}^T \mathbf{q}$, and $|\mathbf{p}|_2$ stands for $(\mathbf{p}^T\mathbf{p})^{\frac{1}{2}}$. The similarity between the words w_1 and w_2 is given by $\mathbf{p_n}(w_1)^T \mathbf{p_n}(w_2)$. We define the variance for a set of k words $\{w_{i_1}, \ldots, w_{i_k}\}$ as follows:

$$\text{var}\,(\{w_{i_1}, \ldots, w_{i_k}\}) = \sum_{j=1}^{k} f_j |\mathbf{p_n}(w_{i_j})|^2 - \left(\sum_{j=1}^{k} f_j \mathbf{p_n}(w_{i_j})\right)^T \left(\sum_{j=1}^{k} f_j \mathbf{p_n}(w_{i_j})\right) \quad (1)$$

where $f_j = \dfrac{1}{k}$, $j = 1, \ldots, k$.

We denote $\displaystyle\sum_{j=1}^{k} f_j \mathbf{p_n}(w_{i_j})$ by $\mathbf{e}\left(\{\mathbf{p_n}(w_{i_1}), \ldots, \mathbf{p_n}(w_{i_k})\}\right)$, and keeping in mind that $\{\mathbf{p_n}(w_{i_j})\}$ are unit vectors rewrite (1) as follows:

$$\mathrm{var}\left(\{w_{i_1}, \ldots, w_{i_k}\}\right) = 1 - \left|\mathbf{e}\left(\{\mathbf{p_n}(w_{i_1}), \ldots, \mathbf{p_n}(w_{i_k})\}\right)\right|^2. \qquad (2)$$

In the next section the variance defined by (2) is used to evaluate quality of same context words clusters.

3   Experiments

The experiments in this section are carried out on four English corpora:

- Tom-"The Adventures of Tom Sawyer" by Mark Twain,
- Yankee-"A Connecticut Yankee in King Arthur's Court" by Mark Twain,
- Karenina-"Anna Karenina" by Leo Tolstoy,
- Reuters-Reuters-21578, distribution 1.0 test collection.

The first three texts are available from http://www.promo.net/pg/, the last collection of 21578 documents is available from David D. Lewis' home page: http://www.research.att.com/ lewis. In addition to the pre–processing described at the beginning of section 2 all hand–indexed entries were removed from Reuters text collection. Only the text between the delimiters <BODY> and </BODY> (excluding the word "Reuters" completing each news item) has been processed. Reuters files with empty text have been removed. The next table provides some statistics on the texts. The last two rows present variance for the four collections and the corresponding cuts (see equation (2)).

corpus	Tom	Yankee	Karenina	Reuters
size (kb)	412	673	2,057	24,000
number of files	1	1	1	19043
unique words	7194	10382	12775	44749
sentences	3709	4802	16996	121696
mean profile size	41	59	78	140
max sentences per word	710	440	1560	14220
MinNumSent	10	30	100	500
cut(MinNumSent) size	487	183	172	467
var(collection)	0.97	0.97	0.96	0.96
var(cut)	0.82	0.63	0.55	0.56

Next we provide a short list of clusters for each text collection. The "var" row indicates the variance of the corresponding cluster.

cluster	Tom	Yankee	Karenina	Reuters
1	trouble wouldn't	head hands	asked answered	dlrs dlr
var	0.196064	0.207273	0.117762	0.0456933
2	reckon won	home heart	conversation chapter	fall decline
var	0.198276	0.219147	0.121811	0.0370496
3	night told	launcelot knight	face eyes	february december
var	0.203541	0.150268	0.123713	0.0333204
4	head looked	master lord	hands hand	increased increase
var	0.207783	0.216585	0.122147	0.0352877
5	harper injun	queen arthur	home day	lower higher
var	0.129439	0.210372	0.127368	0.0468052
6	good thing	sir knights	people man	period months
var	0.199292	0.207819	0.125185	0.0459476
7	face moment	things couldn't	put long	purchase buy
var	0.202919	0.207673	0.13126	0.0441111
8	chapter sawyer	thought mind	smiling smile	reuters reporters
var	0.174602	0.196599	0.124248	0.0272061
9	aunt sid	thousand hundred	turned looked	rose fell
var	0.201479	0.211229	0.124419	0.018597
10	ain't don't	turned moment	vronsky dolly	total estimated
var	0.141682	0.197751	0.125272	0.049047

The following are immediate observations concerning the presented clusters:

1. Better results are generated for larger text collections: the quality of clusters (measured by variance) decreases with increase in collection's size, and, in the opinion of this writer, words in the same cluster become more and more related. Most of our observations are, therefore, related to the Reuters collection.

2. Karenina clusters 4, 8 and Reuters clusters 1, 4 indicate that "same context words" may be used for automatic stemming.

3. It appears that Reuters collection texts contain different "writing styles". For example, in some of the files we observe heavy use of abbreviations. The most striking example is probably the words "dollar" and "dlr". While it is the impression of this author that the words have exactly the

same meaning in the collection, it turns out that their normalized profile vectors are not similar at all. In fact

$$p_n(\text{"dollar"})^T p_n(\text{"dlr"}) = 0.35, \text{ and } p_n(\text{"dollar"})^T p_n(\text{"dlrs"}) = 0.35,$$

while

$$p_n(\text{"dlr"})^T p_n(\text{"dlrs"}) = 0.90, \text{ and } p_n(\text{"dollar"})^T p_n(\text{"currency"}) = 0.74.$$

In fact, the words "dollar" and "currency" belong to the same cluster. We speculate that, as compared with works of Mark Twain and Leo Tolstoy, the Reuters collection contains a mixture of different "writing styles" (abbreviations vs. full words). The clustering approach presented in this paper is unable to select same context words from "mixed jargon" text collections (for additional discussion of related problems see e.g. [8]).

4. Karenina cluster 10 may lead to amusing interpretations. In the next table we display similarity (dot product) between "vronsky" and a number of female personalities.

	anna	betsy	darya	dolly	kitty	lidia	woman	wife	feeling
vronsky	0.677	0.632	0.278	**0.749**	**0.709**	0.267	0.608	**0.703**	0.721

The corresponding table for "anna" is given next

	alexey	husband	konstantin	man	oblonsky	sergey	stepan	vronsky	feeling
anna	0.377	**0.682**	0.235	0.623	0.538	0.179	0.236	**0.677**	0.636

The tables clearly indicate that both, Anna and Vronsky, are no strangers to "feeling."

4 Applications

We believe that the introduced technique can be useful for a number of information retrieval applications. In this section we indicate briefly some preliminary results concerning the following two problems:

1. an automatic corpus dependent stemming,
2. index terms selection.

4.1 Stemming

Following Xu and Croft [20] we refer to corpus–based stemming as generation of word's "equivalence classes to suit the characteristics of a given text corpus." As a preliminary experiment we select words that appear in at least 100 sentences of the Reuters collections. We then select words that start with the letters "cr". The selection contains the following 11 words:

{create, created, credit, creditor, creditors, credits, crisis, crop, crops, crowns, crude}.

The means clustering (see [16]) applied to these words generates the following 6 clusters:
{created, create}, {credits, credit}, {crisis, creditors, creditor}, {crops, crop}, {crude}, {crowns}.

The similarity (dot product) matrix

	create	created	credit	creditor	creditors	credits	crisis	crop	crops	crowns	crude
create	1.00	0.54	0.47	0.33	0.43	0.25	0.42	0.26	0.26	0.34	0.24
created		1.00	0.49	0.26	0.38	0.30	0.34	0.29	0.26	0.35	0.30
credit			1.00	0.45	0.56	0.69	0.40	0.45	0.35	0.59	0.35
creditor				1.00	0.67	0.27	0.58	0.17	0.15	0.28	0.14
creditors					1.00	0.34	0.70	0.22	0.21	0.39	0.22
credits						1.00	0.23	0.32	0.24	0.44	0.27
crisis							1.00	0.24	0.24	0.24	0.22
crop								1.00	0.71	0.40	0.33
crops									1.00	0.28	0.28
crowns										1.00	0.30
crude											1.00

clearly indicates strong similarity between the words

{crisis, creditors, creditor}

in the Reuters collection (the Reuters documents are short business news and words "crisis", "creditors", and "creditor" often occur in the same context). At the same time the results generated by Porter stemming (see [17])

{ creat }, { credit }, { creditor }, { crisi }, { crop }, { crown }, { crude }

miss this connection.

4.2   Index Terms Selection

A basic step involved in the construction of a vector space model is the choice of terms that index documents (see e.g. [4]). If the processing task is to partition a given document collection into clusters of similar documents, then a good choice of index terms is of paramount importance. To provide an example of a "good choice of index terms" consider a set of documents comprised of the following three document collections (available from http://www.cs.utk.edu/ lsi/):

- Medlars Collection (1033 medical abstracts),
- CISI Collection (1460 information science abstracts),
- Cranfield Collection (1398 aerodynamics abstracts).

When all 3891 documents are mixed together into a single collection, and the goal is to partition the collection into three sub-collections to restore documents original membership described above the term "blood" is probably more useful for the task then the term "case". Indeed, while the term "case" occurs in 253 Medlars documents, 72 CISI documents, and 365 Cranfield documents, the term "blood" occurs in 142 Medlars documents, 0 CISI documents, and 0 Cranfield documents. With each term w we associate a three dimensional "direction" vector $\mathbf{d}(w) = (d_1(w), d_2(w), d_3(w))$, so that $d_i(w)$ is the number of documents in collection i containing the term w. So, for example, $\mathbf{d}(\text{"case"}) = (253, 72, 365)$, and $\mathbf{d}(\text{"blood"}) = (142, 0, 0)$. In addition to "blood" terms like "layer" ($\mathbf{d}(\text{"layer"}) = (6, 0, 358)$), or "retriev" ($\mathbf{d}(\text{"retriev"}) = (0, 262, 0)$) are probably much more useful then terms "case", "studi" ($\mathbf{d}(\text{"studi"}) = (356, 341, 238)$), and "found" ($\mathbf{d}(\text{"found"}) = (211, 93, 322)$).

When only the "combined" collection of 3891 documents is available the above described construction of "direction" vectors is not possible. It is of interest to develop algorithms that select "useful" terms when the direction vector $\mathbf{d}(\mathbf{t})$ is not available.

The "profile technique" introduced in this paper allows to define a quality functional $q(w)$ for each term w in the document collection. The quality $q(w)$ is a number between 0 and 1. The larger $q(w)$ is the more "useful" the term w is for the clustering task. While detailed description of the quality functional is beyond the scope of this paper (and will be reported elsewhere), Tables 1 and 2 present 15 "best" and 15 "worst" terms along with their quality scores and direction vectors.

Stemming of the entire collection of 3891 documents produces 16287 unique terms. The results are reported for cut(4) (i.e. for terms that occur in at lest four sentences across the collection). The size of the cut is 5154.

5   Conclusion

Applications of clustering techniques for retrieval systems performance improvement is not new. Already in 1977 in order to enhance the performance of full–text retrieval systems Attar and Fraenkel [3] suggested a procedure based on iterative local–dynamic clustering. At each iteration the system is using retrieved documents to construct "searchonyms". Roughly speaking a word w' is a searchonym of a word w if w' can replace w in the set of retrieved documents. One of the reasons to make searchonyms dependent on a set of retrieved documents (rather than on the entire collection) are computing limitations of 1977.

The same context words introduced in the paper are constructed based on the similarity measure provided by the words context throughout the entire text collection. Selection of same context words requires clustering of large

term	quality(term)	d_1(term)	d_2(term)	d_3(term)
lenticular	0.186	2	0	0
layer	0.186	6	0	358
boundari	0.178	0	7	413
aortic	0.176	26	0	0
inform	0.174	28	614	44
ventricular	0.174	53	0	0
laminar	0.170	0	0	231
estron	0.168	4	0	0
number	0.168	92	204	568
retriev	0.165	0	262	0
shell	0.164	0	4	105
septal	0.163	25	0	0
blunt	0.162	1	0	119
nadh2	0.162	2	0	0
axial	0.162	0	0	136

Table 1. 15 "best" terms in slice(4)

term	quality(term)	d_1(term)	d_2(term)	d_3(term)
present	0.004	236	314	506
includ	0.004	75	169	225
experi	0.004	105	133	152
work	0.004	17	245	112
show	0.004	168	97	202
shown	0.004	58	62	285
oper	0.004	70	184	67
found	0.003	211	93	322
determin	0.003	108	116	299
larg	0.003	80	175	201
gener	0.003	76	311	329
discuss	0.003	142	262	271
studi	0.002	356	341	238
develop	0.002	176	366	264
case	0.002	253	72	365

Table 2. 15 "worst" terms in slice(4)

data sets. While a number of clustering algorithms capable of dealing with large data collections has been currently reported in the literature (see e.g. [2], [6], [10], [15], and [21]), the choice of appropriate clustering algorithms along with many other technical details (among them appropriate definitions of the profile vectors, and possible choices of the metric) require further research.

It is hoped that the introduced technique when combined with existing information retrieval tools will provide a useful automatic corpus dependent text analysis tool. In particular, we feel that separate construction of same context words clusters for nouns, verbs, adjectives, and adverbs has a potential further enhance dimensionality reduction. Hence, a combination of the same context words technique with, for example, Princeton developed Word-Net may lead to an efficient choice of words for information retrieval related problems (for detailed discussion of the issue we refer the reader to [12]).

References

1. Agrawal, R., Imielinski, T., Swami, A. (1993). Mining association rules between sets of items in large databases. Proc. ACM SIGMOD, 207–216
2. Alsabti, K., Ranka, S., Singh, V. (1999). An efficient space–partitioning based algorithm for the k-means clustering. PAKDD, 355–359
3. Attar, R., Fraenkel, A.S. (1977). Local feedback in full–text retrieval systems. J. Assoc. Comput. Mach. **24**, 397–417
4. Berry, M., Browne, M. (1999). *Understanding Search Engines*. SIAM, Philadelphia
5. Berry, M., Drmac, Z., Jessup, E.R. (1999). Matrices, vectors spaces, and information retrieval. SIAM Rev. **41**, 335–362
6. Boley, D. (1998). Principal directions divisive partitioning. Data Min. Knowl. Disc. **2**, 325-344
7. Bottou, L., Bengio, Y. (1995). Convergence properties of the k–means algorithms. In: *Advances in Neutral Information Processing Systems 7*, Tesario, G., Touretzky, D. (Eds.), The MIT Press, Massachusets, 585–592
8. Castellanos, M., Stinger, J.R. (2001). A practical approach to extracting relevant sentences in the presence of dirty text. In: *Workshop on Text Mining*. Berry, M.W. (Ed.), Chicago, Illinois, 15–22
9. Deerwester, S., Dumas, S., Furnas, G., Landauer, T., (1990). Indexing by Latent Semantic Analysis. J. Am. Soc. Inform. Sci. **41**, 391–407
10. Dhillon, I.S., Modha, D.S. (2000). A data–clustering algorithm on distributed memory multiprocessors, in Large-Scale Parallel Data Mining. Lect. Notes Artif. Int. **1759**, 245–260
11. Grefenstette, G. (1994). *Explorations in Automatic Thesaurus Discovery*. Kluwer, Boston
12. Jing, Y., Croft, W.B. (1994). An association thesaurus for information retrieval. In: *Proceedings of RIAO 94*, 146–160
13. Kleinberg, J., Tomkins A. (1999). Applications of linear algebra in information retrieval and hypertext analysis. In: *Proceedings of the eighteenth ACM SIGMOD-SIGACT-SIGART symposium on Principles of database systems*, 185–193
14. Kowalski, G. (1997). *Information Retrieval Systems*. Kluwer, Boston
15. Kogan, J. (2001). Clustering large unstructured document sets. In: *Computational Information Retrieval*. Berry, M.W. (Ed.), SIAM, Philadelphia, 115–125
16. Kogan, J. (2001). Means clustering for text data. In: *Workshop on Text Mining*. Berry, M.W. (Ed.), Chicago, Illinois, 47–54

17. Porter, M.F. (1980). An algorithm for suffix stripping. Program **14**, 130–137
18. Schütze, H., Pedersen, J.O. (1995). Information retrieval based on word senses. In: *Proceedings of the Symposium on Document Analysis and Information Retrieval* **4**, 161–175
19. Selim, S.Z., Ismail, M.A. (1984). K–means–type algorithms: a generalized convergence theorem and characterization of local optimality. IEEE T. Pattern Anal. **6**, 81–87
20. Xu, J., Croft, W.B. (1998). Corpus-based stemming using co–occurrence of word variance. ACM T. Inform. Syst. **16**, 61–81
21. Zhang, T., Ramakrishnan, R., Livny, M. (1996). BIRCH: an efficient data clustering method for very large databases. *Proceedings of the 1996 ACM SIGMOD International Conference on Management of Data*, 103–114

Studying Treatment Response to Inform Treatment Choice

Charles F. Manski*

Department of Economics and Institute for Policy Research Northwestern University, Evanston, Illinois, USA

Abstract. An important practical objective of empirical studies of treatment response is to provide decision makers with information useful in choosing treatments. Often the decision maker is a planner who must choose treatments for the members of a heterogeneous population; for example, a physician may choose medical treatments for a population of patients. Studies of treatment response cannot provide all the information that planners would like to have as they choose treatments, but researchers can be of service by addressing several questions: How should studies be designed in order to be most informative? How should studies report their findings so as to be most useful in decision making? How should planners utilize the information that studies provide? This paper addresses aspects of these broad questions, focusing on pervasive problems of identification that arise when studying treatment response and making treatment choices.

1 Introduction

An important practical objective of empirical studies of treatment response is to provide decision makers with information useful in choosing treatments. Often the decision maker is a planner who must chooses treatments for a heterogeneous population. The planner might, for example, be a physician choosing medical treatments for a population of patients. Physicians use findings of medical research to evaluate the merits of alternative treatment rules.

* This research was supported in part by National Science Foundation grant SES-0001436. I have benefitted from the opportunity to present aspects of this work at the December, 2000 Conference on Foundations of Statistical Inference in Shoresh, Israel; the September 2000 Symposium on Causal Inference at the Johns Hopkins School of Hygiene and Public Health; and in seminars at Eli Lilly & Co., New York University, Northwestern University, the University of Bristol, the University of Chicago, and the University of Virginia. I have also received valuable comments from John Pepper. The author thanks William G. Henderson, Domenic Reda, and David Williams of the Edward Hines, Jr. Hospital, U.S. Department of Veterans Affairs Cooperative Studies Program Coordinating Center, Hines, Illinois for providing and helping to interpret the randomized-clinical-trial data analyzed in Section 3. The Department of Veterans Affairs retains ownership of the data. The views expressed in this paper are those of the author and are not necessarily endorsed by the Department of Veterans Affairs.

It is unrealistic to think that studies of treatment response can provide all the information that planners would like to have as they choose treatments. However, researchers can aim to improve treatment choice by addressing several questions: How should studies be designed in order to be most informative? How should studies report their findings so as to be most useful in decision making? How should planners utilize the information that studies provide? This paper draws on the author's research to address aspects of these broad questions.

My starting point is the decision theoretic framework of [16] and [17]. This assumes that the planner observes some covariates for each member of the population-to-be-treated; for example, a physician may observe a patient's demographic attributes, medical history, and the results of diagnostic tests. The observed covariates determine the set of treatment rules that are feasible for the planner to implement: the set of feasible rules is the set of all functions mapping the observed covariates into treatments. Each member of the population has a response function which maps treatments into a real-valued outcome of interest; perhaps a measure of health status in the case of medical treatment. I assume that the planner wants to choose a treatment rule that maximizes the population mean outcome; in economic terms, the planner wants to maximize a utilitarian social welfare function. Under these assumptions, an optimal treatment rule assigns to each member of the population a treatment that maximizes mean outcome conditional on the person's observed covariates. Hence studies of treatment response are useful to the degree that they enable the planner to learn how mean outcomes vary with treatments and covariates.

Section 2 formalizes these ideas, from which I conclude that heterogeneity in treatment response should be a central concern in study design. Researchers should bear in mind the planner's problem when deciding what population to study and what covariate information to report on study subjects. I reconsider the widely held view that studies of treatment response should be judged primarily by their internal validity and only secondarily by their external validity.

Section 3 examines the implications for treatment choice of pervasive identification problems in studies of treatment response. I first explain in general terms how identification problems in the empirical analysis of treatment response generate ambiguity about the identity of optimal treatment choices. Then, drawing on [8], I examine the particular identification problem created by missing outcome and covariate data and use a randomized clinical trial (RCT) of treatments for hypertension (see [22]) to illustrate findings.

Studies of treatment response generally report outcomes for finite samples of subjects, not for entire study populations. Hence planners wanting to use study findings not only face identification problems but also must perform statistical inference from the sample to the population. This short article abstracts from the problem of statistical inference. However, I do address this important matter in [15].

2   Treatment Choice in Heterogeneous Populations

Section 2.1 formalizes the planner's problem as in Manski [16] and [17]. With this background, Sects. 2.2 and 2.3 draw implications for studies of treatment response.

2.1   The Planner's Problem

I suppose that there is a finite set T of mutually exclusive and exhaustive treatments. A planner must choose a treatment rule assigning a treatment in T to each member of a population J. Each person $j \in J$ has a *response function* $y_j(\cdot) : T \to Y$ mapping treatments into real-valued outcomes $y_j(t) \in Y$. A *treatment rule* is a function $\tau(\cdot) : J \to T$ specifying which treatment each person is assigned. Thus person j's outcome under rule $\tau(\cdot)$ is $y_j[\tau(j)]$. This notation maintains the assumption of individualistic treatment made commonly in analyses of treatment response; that is, a person's outcome may depend on the treatment he is assigned, but not on the treatments assigned to others.

The planner is concerned with the distribution of outcomes across the population, not with the outcomes of particular persons. Hence it is convenient to let the population be a probability space, say (J, Ω, P), with Ω the σ-algebra and P the probability measure. Now the population mean outcome, or *social welfare*, under treatment rule $\tau(\cdot)$ is well-defined as

$$E\{y_j[\tau(j)]\} \equiv \int y_j[\tau(j)]dP(j). \tag{1}$$

I assume that the planner wants to choose a treatment rule that maximizes $E\{y_j[\tau(j)]\}$. This criterion function has normative, analytical, and practical appeal. Maximization of a population mean outcome, or perhaps some weighted average outcome, is the standard normative criterion of the public economics literature on social planning; the outcome of interest measures the social benefits minus costs of a treatment. The linearity of the expectation operator yields substantial analytical simplifications, particularly through use of the law of iterated expectations. The practical appeal is that a planner choosing treatments to maximize the mean population outcome will want to learn average treatment effects, the dominant form of treatment effect reported in the empirical literature on treatment response. Other criterion functions generate interest in other forms of treatment effect.

The planner observes certain covariates $x_j \in X$ for each member of the population. The planner cannot distinguish among persons with the same observed covariates and so cannot implement treatment rules that systematically differentiate among these persons. Hence the feasible non-randomized rules are functions mapping the observed covariates into treatments.

To formalize the planner's problem, let Z denote the space of all functions mapping X into T. Let $z(\cdot) \in Z$. Then the feasible treatment rules have the form

$$\tau(j) = z(x_j), \qquad j \in J. \tag{2}$$

Let $P[y(\cdot), x]$ be the probability measure on $Y^T \times X$ induced by $P(j)$. Let $E\{y[z(x)]\} \equiv \int y[z(x)]dP[y(\cdot), x]$ denote the expected value of $y[z(x)]$. Then the planner wants to solve the problem

$$\max_{z(\cdot) \in Z} E\{y[z(x)]\}. \tag{3}$$

In practice, institutional constraints may restrict the feasible treatment rules to a proper subset of Z. For example, the planner may be precluded from using certain covariates (say race or gender) to assign treatments. The analysis in this paper continues to hold if x is defined to be the covariates that the planner is permitted to consider, rather than the full vector of covariates that the planner observes.

It is easy to show that the solution to the planner's problem is to assign to each member of the population a treatment that maximizes mean outcome conditional on the person's observed covariates. Let $1[\cdot]$ be the indicator function taking the value one if the logical condition in the brackets holds and the value zero otherwise. For each $z(\cdot) \in Z$, use the law of iterated expectations to write

$$E\{y[z(x)]\} = E\{E\{y[z(x)]|x\}\} = E\{\Sigma_{t \in T} E[y(t)|x] \cdot 1[z(x) = t]\} \tag{4}$$

$$= \int \Sigma_{t \in T} E[y(t)|x] \cdot 1[z(x) = t]dP(x).$$

For each $x \in X$, the integrand $\Sigma_{t \in T} E[y(t)|x] \cdot 1[z(x) = t]$ is maximized by selecting $z(x)$ to maximize $E[y(t)|x]$ on $t \in T$. Hence rule $z^*(\cdot)$ is optimal if, for $x \in X$, $z^*(x)$ solves the problem $\max_{t \in T} E[y(t)|x]$. The optimized population mean outcome is $E\{\max_{t \in T} E[y(t)|x]\}$.

The set of feasible treatment rules grows as more covariates are observed. Hence the optimal mean outcome achievable by the planner cannot fall, and may rise, as more covariates are observed. The value of covariate information is appropriately measured by the difference between the optimal mean outcome achievable with and without use of this information. This is

$$V(X) \equiv E\{\max_{t \in T} E[y(t)|x]\} - \max_{t \in T} E[y(t)]. \tag{5}$$

Inspection of (5) shows that covariate information has no value if there exists a common optimal treatment; that is, a $t^* \in T$ such that $z^*(x) = t^*$, almost everywhere on X. Covariate information does have value if optimal treatments vary with x.

More generally, we may compare the value of observing distinct covariate vectors, say x and w. A planner who knows the conditional mean treatment responses $E[y(\cdot)|x]$ and $E[y(\cdot)|w]$ should prefer observation of x to w if and only if $E\{\max_{t\in T} E[y(t)|x]\} \geq E\{\max_{t\in T} E[y(t)|w]\}$. This criterion for comparison of x and w differs from the prediction criterion familiar in statistical decision theory. The prediction criterion supposes that, for each $t \in T$, one wants to predict $y(t)$ as well as possible in the sense of minimizing expected square loss. The best predictors conditional on x and w are $E[y(t)|x]$ and $E[y(t)|w]$ respectively. A statistician who knows $E[y(t)|x]$ and $E[y(t)|w]$ and wants to predict $y(t)$ as well as possible should prefer x to w if and only if $E\{y(t) - E[y(t)|x]\}^2 \leq E\{y(t) - E[y(t)|w]\}^2$.

2.2   Reporting Covariate Information in Studies of Treatment Response

Researchers should bear in mind the treatment choice problem when deciding what covariate information to report on study subjects. Yet there is often a wide disparity between the covariates that planners can observe and the covariate information reported in studies of treatment response. For example, physicians commonly observe medical histories, diagnostic test findings, and demographic attributes for the patients that they treat. Yet the journal articles that report on RCTs often provide scant covariate information for study subjects, describing outcomes only within broad risk-factor groups.

There seem to be several reasons why studies of treatment response report little covariate information. (I say "seem to" because these reasons are rarely stated explicitly.) Some researchers may assume that there exists a common optimal treatment across the population of interest; then covariate information has no value. Concern for the confidentiality of subjects' identities may inhibit reporting covariate data. Editorial restrictions on the lengths of journal articles may prevent researchers from reporting useful findings. Sampling variability may inhibit researchers from reporting treatment response conditional on covariates. In particular, findings may be reported only if they meet conventional criteria for statistical precision. Whenever there is reason to think that treatment response may vary with covariates that planners can observe, researchers should aim to report findings on mean treatment response conditional on these covariates. Subject to considerations of subject confidentiality and space constraints, research journals should encourage publication of such findings. When journal space constraints prevent publication of useful findings, researchers should make them available on the web or through other means.

2.3   The Study Population and the Population-to-be-treated

A longstanding issue in study design concerns the importance of correspondence between the study population and the population-to-be-treated. This matter was downplayed in the influential work of Donald Campbell, who argued that studies of treatment effects should be judged primarily by their internal validity and only secondarily by their external validity (see [3], [4]). Campbell's view has recently been endorsed by Rosenbaum [25], who recommends that observational studies of human subjects aim to approximate the conditions of laboratory experiments. Rosenbaum, like Campbell, downplays the importance of having the study population be similar to the population of interest, writing (page 259): "Studies of samples that are representative of populations may be quite useful in describing those populations, but may be ill-suited to inferences about treatment effects".

From the perspective of treatment choice, the Campbell-Rosenbaum position is well grounded if treatment response is homogeneous. Then researchers can aim to learn about treatment response in easy-to-analyze study populations and planners can be confident that research findings can be extrapolated to populations of interest. In human populations, however, homogeneity of treatment response may be the exception rather than the rule. Whether the context be medical or educational or social, there is often reason to think that people vary in their response to treatment. To the degree that treatment response is heterogeneous, a planner cannot readily extrapolate research findings from a study population to a population of interest, as optimal treatments in the two may differ. Hence correspondence between the study population and the population-to-be-treated assumes considerable importance.

When the objective is to inform treatment choice in heterogeneous populations, I see no reason to give internal validity primacy relative to external validity. To be fair, researchers who stress internal validity may have objectives other than to inform treatment choice. For example, Angrist, Imbens, Rubin in [1] state their goal to be the discovery of "causal effects", without reference to a treatment-choice problem.

3   Identification Problems and Treatment Choice under Ambiguity

Ideally, a planner facing the treatment choice problem described in Sect. 2 would like studies of treatment response to reveal in full how mean outcomes vary with treatments and covariates. In practice, problems of identification and statistical inference limit the information that studies can provide. Statistical and identification problems are logically distinct, and it is analytically useful to consider them sequentially. Here I suppose that researchers are able to draw random samples of unlimited size from their study populations and hence know (almost surely) whatever population features their sampling processes are capable of revealing. In Manski [15], I suppose that researchers are only able to draw random samples of finite size and hence must make statistical inferences about their study populations. Section 3.1, drawing again on [16], [17], continues the formalization of the planner's problem begun in Sect. 2.1 and describes in general terms how identification problems generate ambiguity about the identity of optimal treatment choices. Section 3.2 discusses the specific identification problem that arises in studies with missing outcome and covariate data. Section 3.3 illustrates with data from an RCT of treatments for hypertension.

3.1 The Planner's Problem, Continued

By equations (3) and (4), the planner would like to choose a treatment rule that solves the problem

$$\max_{z(\cdot) \in Z} \int \Sigma_{t \in T} E[y(t)|x] \cdot 1[z(x) = t] dP(x). \tag{6}$$

The covariates x are observable, so it is realistic to assume that the planner can learn the distribution $P(x)$ of covariates in the population-to-be-treated. Research on treatment response is motivated by the planner's desire to learn $\{E[y(\cdot)|x]\}, x \in X\}$.

Identification problems limit the information that studies of treatment response provide. Considering the matter in abstraction, suppose a planner learns from the available studies that mean treatment response conditional on the observed covariates lies in some *identification region H*; that is, $\{E[y(\cdot)|x], x \in X\} \in H$, for some $H \subset Y^T \times X$. This information may not suffice to solve problem (6), in which case the planner faces a problem of treatment choice under *ambiguity*.

What should a planner do in such a situation? Clearly he should not choose a dominated treatment rule: a rule $z(\cdot)$ is dominated if there exists another feasible rule, say $z'(\cdot)$, which necessarily yields at least the social welfare of $z(\cdot)$ and which performs strictly better than $z(\cdot)$ in some state of nature. Thus, $z(\cdot)$ is dominated if there exists a $z'(\cdot) \in Z$ such that

$$\int \sum_{t \in T} \eta(t, x) \cdot 1[z(x) = t] dP(x) \tag{7a}$$

$$\leq \int \sum_{t \in T} \eta(t, x) \cdot 1[z'(x) = t] dP(x), \qquad \forall \eta \in H$$

$$\text{and} \int \sum_{t \in T} \eta(t, x) \cdot 1[z(x) = t] dP(x) \tag{7b}$$

$$< \int \sum_{t \in T} \eta(t, x) \cdot 1[z'(x) = t] dP(x), \text{some } \eta \in H,$$

where $[\eta(\cdot, x), x \in X]$ denotes a feasible value of $\{E[y(\cdot)|x], x \in X\}$. The central difficulty of treatment choice under ambiguity is that there is no clearly best way to choose among undominated treatment rules. The most that can be said is that decision theory suggests a variety of "reasonable" procedures, including the maximin rule and Bayes rules.

An unfortunate characteristic of empirical research on treatment response has been that it gives planners little sense of how identification problems limit inference. Researchers commonly report point estimates of mean treatment response, not estimates of identification regions. The reported estimates often have fragile foundations, as becomes plain from observing the persistent disagreements among researchers about the credibility of alternative identifying assumptions. I have long argued that researchers and planners alike would be better served if the customary practice were to first report the limited inferences that are possible using only knowledge of the sampling process generating the data, and then report tighter inferences that combine the available data with highly credible assumptions about treatment selection and response. See [11], [13], [18], and [19].

3.2 Treatment Choice Using Studies with Missing Outcome and Covariate Data

A particularly important source of incomplete identification is missing data, which afflicts every study of treatment response in one way or another. All studies have missing outcome data due to the fact that counterfactual (aka latent or potential) outcomes are unobservable - at most one can observe the outcomes that persons experience under the treatments that they actually receive. Outcome data may also be missing due to attrition of subjects from randomized trials or due to nonresponse in observational studies.

The specific form of H implied by missing outcome data depends on the prior information that one can combine with the available empirical evidence. The classical assumption that outcome data are missing completely at random (MCAR) implies that H is a point; in this best case scenario, an analyst can simply ignore sample realizations with missing data. Intent-to-treat (ITT) analysis of RCTs with noncompliance is based on another best-case scenario: one assumes that the compliance behavior of subjects in the trial correctly predicts the compliance behavior that would occur when treatments are assigned in practice. [11], [12] characterized the worst-case scenario in which one has no prior information about the missing data process. Middle-ground cases which bring to bear some prior information but not enough to reduce H to a point have been studied in [2], [9], [12], [13], [14], [19], [20], and [23].

Missing covariate data is a common occurrence in studies of treatment response, but has received much less research attention than has missing outcome data. [10], [24], and [26] pose best-case scenarios asserting enough prior information to achieve point identification. The worst-case scenario was first studied in [7]. For any specified value of x, this article gives sharp bounds on $E[y(\cdot)|x]$ in two observational settings: only covariate data are missing, and (covariate, outcome) data are jointly missing.

Horowitz and Manski in [8] have analyzed identification of mean treatment response when outcome and/or covariate data may be missing. We suppose that outcomes are binary and derive sharp bounds on $E[y(\cdot)|x]$ in two informational settings – the worst-case scenario (Theorem 1) and the partial information setting where it is known that covariate data are MCAR (Theorem 3). These theorems apply to general missing data problems – some observations may be complete, some may have missing outcome data, others may have missing covariate data, and still others may have jointly missing (covariate, outcome) data. I illustrate below.

3.3   Choosing Treatments for Hypertension Using Data from a Trial with Missing Data

Physicians routinely choose treatments for hypertension. Medical research has sought to provide guidance through the conduct of RCTs comparing alternative treatments. Such trials inevitably have missing data. I illustrate here how physicians might use the data from a recent trial to inform treatment choice, without imposing untenable assumptions about the distribution of the missing data.

Materson et al. in [21] presented findings from a RCT of treatments for hypertension sponsored by the U.S. Department of Veteran Affairs (DVA). Male veteran patients at 15 DVA hospitals were randomly assigned to one of 6 antihypertensive drug treatments or to placebo: hydrochlorothiazide ($t = 1$), atenolol ($t = 2$), captopril ($t = 3$), clonidine ($t = 4$), diltiazem ($t = 5$), prazosin ($t = 6$), placebo ($t = 7$). The trial had two phases. In the first, the dosage that brought diastolic blood pressure (DBP) below 90 mm Hg was determined. In the second, it was determined whether DBP could be kept below 95 mm Hg for a long time. Treatment was defined to be successful if DBP < 90 mm Hg on two consecutive measurement occasions in the first phase and DBP $\leq$ 95 mm Hg in the second. Treatment was deemed unsuccessful otherwise. Thus the outcome of interest was binary, with $y = 1$ if the criterion for success is met and $y = 0$ otherwise. [21] recommended that physicians making treatment choices should consider this medical outcome variable as well as patient's quality of life and the cost of treatment.

The [21] article examined how treatment response varies with the race and age of the patient. There were no missing data on the race and age covariates. The authors performed an ITT analysis that interpreted attrition from the trial as lack of success; from this perspective there were no missing outcome data either. Horowitz and Manski in [8] obtained the trial data and used them to examine how treatment response varies with another covariate that does have missing data. This was the biochemical indicator "renin response", taking the values $x =$(low, medium, high), which had previously been studied as a factor that might be related to successful treatment (see [6]). Renin-response was measured at the time of randomization, but data were missing for some subjects in the trial. Horowitz and Manski also stepped back from the ITT interpretation of attrition as lack of success. Instead, we viewed subjects who leave the trial as having missing outcome data. The pattern of missing covariate and outcome data is shown in [8], Table 1, reproduced here.

Table 1. Missing data in the DVA hypertension trial

Treat-ment	Number Randomized	Observed Successes	None Missing	Missing Only y	Missing Only x	Missing (y, x)
1	188	100	173	4	11	0
2	178	106	158	11	9	0
3	188	96	169	6	13	0
4	178	110	159	5	13	1
5	185	130	164	6	14	1
6	188	97	164	12	10	2
7	187	57	178	3	6	0

Horowitz and Manski in [8] used their Theorems 1 and 3 to estimate sharp bounds on the success probabilities $\{P[y(t) = 1|x], t = 1, \ldots, 7\}$, first without imposing assumptions on the distribution of missing data and then assuming that missing covariate data are MCAR. Rather than report the bounds on the success probabilities directly, the article reported the implied bounds on the average treatment effects $\{P[y(t) = 1|x] - P[y(7) = 1|x], t = 1, \ldots, 6\}$, which measure the efficacy of each treatment relative to the placebo. This reporting decision was motivated by the traditional research problem of testing the hypothesis of zero treatment effect. The problem of treatment choice was not explicitly examined.

Table 2 reports the estimates of the worst-case bounds on the success probabilities. To keep attention focused on the identification problem, suppose that the estimates are the actual bounds rather than finite-sample estimates. Consider a physician who accepts the DVA success criterion, observes renin response, and has no prior information on mean treatment response or the distribution of missing data. How might this physician choose treatments in a population analogous to that studied in the DVA trial?

First, the physician should eliminate the dominated treatments. For patients with low renin response, treatments 1, 2, 3, 4, 6, and 7 are all dominated by treatment 5, which has the greatest lower bound (.66). For patients with medium renin response, treatments 1, 3, 6, and 7 are dominated by treatment 5, which again has the greatest lowest bound (.68). For patients with high renin response, treatments 1, 6, and 7 are dominated by treatment 2, which has the greatest lowest bound (.64). Thus, without imposing any assumptions on the distribution of missing data, the physician can reject treatments 1, 6, and 7 for all patients, can reject treatment 3 for patients with medium renin

Table 2. Worst-case bounds on success probabilities conditional on renin response

Renin Reponse	Treatment						
	1	2	3	4	5	6	7
Low	[.54, .61]	[.52, .62]	[.43, .53]	[.58, .66]	[.66, .76]	[.54, .65]	[.29, .32]
Medium	[.47, .62]	[.60, .74]	[.53, .68]	[.50, .69]	[.68, .85]	[.41, .65]	[.27, .32]
High	[.28, .50]	[.64, .86]	[.56, .75]	[.63, .84]	[.55, .78]	[.34, .59]	[.28, .40]

response, and can determine that treatment 5 is optimal for patients with low renin response.

In the absence of assumptions about the distribution of missing data, there is no single "right" way for the physician to choose among undominated treatments for patients with medium and high renin response. A physician using the maximin rule would choose treatment 5 for patients with medium renin response and treatment 2 for patients with high renin response. This is a reasonable treatment rule, but one cannot say that it is an optimal rule.

Exploring the reasons for missing data in the DVA trial, Horowitz and Manski in [8] did not find a credible basis to impose assumptions on the distribution of missing outcome data, but did find it plausible to assume that missing covariate data are MCAR. This assumption generates tighter bounds on mean treatment response. Table 3 presents the resulting estimates of bounds on success probabilities.

Table 3. MCAR-covariates bounds on success probabilities conditional on renin response

Renin Reponse	Treatment						
	1	2	3	4	5	6	7
Low	[.57, .58]	[.54, .60]	[.44, .49]	[.61, .63]	[.69, .74]	[.56, .62]	[.31, .32]
Medium	[.52, .57]	[.66, .71]	[.59, .59]	[.55, .63]	[.81, .81]	[.46, .57]	[.32, .32]
High	[.35, .35]	[.75, .83]	[.65, .65]	[.77, .77]	[.67, .70]	[.40, .47]	[.33, .40]

These tighter bounds resolve most of the ambiguity in treatment choice. A physician who accepts the assumption that covariate data are MCAR can conclude that treatment 5 is optimal for patients with low and medium renin response. This physician can narrow consideration to treatments 2 and 4 for patients with high renin response, but the data combined with the MCAR assumption do not suffice to choose between these two treatments.

4 Conclusion

The objective of informing treatment choice provides an explicit practical motivation for empirical study of treatment response. The treatment-choice perspective systematically affects how one should select a study population and cope with identification problems. It also systematically affects how one should make use of finite-sample data (see [15]).

I would particularly stress that empirical research seeking to inform treatment choices differs from analyses that aim to perform classical hypothesis tests. Empirical research on treatment response has been strongly influenced by the classical theory of hypothesis testing, especially by the idea of testing the null hypothesis of zero average treatment effect in the study population. This hypothesis is institutionalized in the Food and Drug Administration drug approval process, which calls for comparison of a treatment under study $(t = 1)$ with a placebo or an approved treatment $(t = 0)$. Approval of treatment 1 normally requires rejection of the null hypothesis of zero average treatment effect $\{H_0 : E[y(1)] = E[y(0)]\}$ in two independent RCTs (see [5]). The null hypothesis of zero treatment effect is prominent in experimental design, as researchers use norms for statistical power to choose sample sizes. Moreover, when studies are performed, findings may go unreported or may be deemed to be "insignificant" if they do not meet test-based criteria for statistical precision. It would be of much interest to reconsider the present FDA drug approval process and current norms for experimental design from the treatment-choice perspective.

References

1. Angrist, J., Imbens, G., Rubin, D. (1996). Identification of Causal Effects Using Instrumental Variables. J. Am. Stat. Assoc. **91**, 444–455
2. Balke, A., Pearl, J. (1997). Bounds on Treatment Effects from Studies With Imperfect Compliance. J. Am. Stat. Assoc. **92**, 1171–1177
3. Campbell, D. (1984). Can We Be Scientific in Applied Social Science?. Evaluation Studies Review Annual. **9**, 26–48
4. Campbell, D., Stanley, R. (1963). *Experimental and Quasi-Experimental Designs for Research.* Rand McNally, Chicago
5. Fisher, L., Moyé, L. (1999). Carvedilol and the Food and Drug Administration Approval Process: An Introduction. Control. Clin. Trials. **20**, 1–15
6. Freis, E., Materson, B., Flamenbaum,W. (1983).Comparison of Propranolol or Hydorchlorothiazide Alone for Treatment of Hypertension, III: Evaluation of the Renin-Angiotensin System. Am. J. Med. **74**, 1029–1041
7. Horowitz, J., Manski, C. (1998). Censoring of Outcomes and Regressors Due to Survey Nonresponse: Identification and Estimation Using Weights and Imputations. J. Econometrics **84**, 37–58
8. Horowitz, J., Manski, C. (2000). Nonparametric Analysis of Randomized Experiments with Missing Covariate and Outcome Data. J. Am. Stat. Assoc. **95**, 77–84

9. Hotz, J., Mullins, C., Sanders, S. (1997). Bounding Causal Effects Using Data from a Contaminated Natural Experiment: Analyzing the Effects of Teenage Childbearing. Rev. Econ. Stud. **64**, 575–603

10. Little, R. (1992). Regression with Missing X's: A Review. J. Am. Stat. Assoc. **87**, 1227–1237

11. Manski, C. (1989). Anatomy of the Selection Problem. J. Hum. Resour. **24**, 343–360

12. Manski, C. (1990). Nonparametric Bounds on Treatment Effects. Am. Econ. Rev. **80**, 319–323

13. Manski, C. (1995). *Identification Problems in the Social Sciences.* Harvard University Press, Cambridge

14. Manski, C. (1997). Monotone Treatment Response. Econometrica **65**, 1311–1334

15. Manski, C. (1999). Statistical Treatment Rules for Heterogeneous Populations. National Bureau of Economic Research Technical Working Paper No. 242

16. Manski, C. (2000). Identification Problems and Decisions Under Ambiguity: Empirical Analysis of Treatment Response and Normative Analysis of Treatment Choice. J. Econometrics **95**, 415–442

17. Manski, C. (2002). Treatment Choice Under Ambiguity Induced by Inferential Problems. J. of Statistical Planning and Inference **105**, 67–82

18. Manski, C., Sandefur, G., McLanahan, S., Powers, D. (1992). Alternative Estimates of the Effect of Family Structure During Adolescence on High School Graduation. J. Am. Stat. Assoc. **87**, 25–37

19. Manski, C., Nagin, D. (1998). Bounding Disagreements About Treatment Effects: A Case Study of Sentencing and Recidivism. Sociol Methodol **28**, 99–137

20. Manski, C., Pepper, J. (2000). Monotone Instrumental Variables: With an Application to the Returns to Schooling. Econometrica **68**, 997–1010

21. Materson, B., Reda, D., Cushman, W., Massie, B., Freis, E., Kochar, M., Hamburger, R., Fye, C., Lakshman, R., Gottdiener, J., Ramirez, E., Henderson, W. (1993). Single-Drug Therapy for Hypertension in Men: A Comparison of Six Antihypertensive Agents with Placebo. New Engl. J. Med. **328**, 914–921

22. Materson, B., Reda, D., Cushman, W. (1995). Department of Veterans Affairs Single-Drug Therapy of Hypertension Study: Revised Figures and New Data. Am. J. Hypertens **8**, 189–192

23. Robins, J. (1989). The Analysis of Randomized and Non-Randomized AIDS Treatment Trials Using a New Approach to Causal Inference in Longitudinal Studies. In: Sechrest, L., Freeman, H., Mulley,A. (Eds.) *Health Service Research Methodology: A Focus on AIDS.* NCHSR, U.S. Public Health Service.

24. Robins, J., Rotnitzky, A., Zhao, L. (1994). Estimation of Regression Coefficients When Some Regressors Are Not Always Observed. J. Am. Stat. Assoc. **89**, 846–866

25. Rosenbaum, P. (1999). Choice as an Alternative to Control in Observational Studies. Stat. Sci. **14**, 259–304

26. Wang, C., Wang, S., Zhao, L., Ou, S. (1997). Weighted Semiparametric Estimation in Regression Analysis with Missing Covariate Data. J. Am. Stat. Assoc. **92**, 512–525

Part II

Bayesian Methods and Modelling

Some Interactive Decision Problems Emerging in Statistical Games

Bruno Bassan[1], Marco Scarsini[2] and Shmuel Zamir[3]

[1] Department of Mathematics, University of Rome 1, Italy
[2] Department of Statistics, University of Turin, Italy
[3] Center for Rationality, Hebrew University of Jerusalem, Israel

Abstract. We consider games which arise when two statisticians must make a decision simultaneously, and the loss function depends on both decisions. We are interested, in particular, in situations when information is detrimental, in a sense to be made precise. We show that in certain problems related to Bayesian testing and prediction the phenomenon of information rejection occurs for certain values of the parameters involved.

1 Introduction

In this note we expand on previous work by the same authors [4] concerning the possibility that two interacting statisticians might prefer to refuse free information. This phenomenon of *information rejection* may occur when the loss of a statistician depends not only on his action and on the state of Nature, but also on the decision made by another statistician. We refer to [4] for general considerations on the problem, and also for references on the relation between statistics and game theory.

Some real situations fit into the scheme of interacting statisticians. For example, the so-called "inspection games", where the statistician of the inspected party is trying to cheat the inspecting colleague (see [1]). Think also of a buyer and a seller simultaneously testing a sample each from a stock of items.

We consider here two examples that were presented in [4], drawn from the theory of Bayesian testing and Bayesian prediction, and we rephrase them in greater generality.

2 Two Interacting Statisticians and Information Refusal

We shall consider two examples, relevant in statistics, of games in which information rejection occurs, in the sense specified below. We shall use the terms "player" and "statistician" indifferently.

The games considered are as follows:

1. Nature chooses between the following bimatrices of payoffs:

$$G_A: \quad \begin{array}{c|c|c|} & a_1 & a_0 \\ \hline a_1 & 0,0 & 0,0 \\ \hline a_0 & 0,0 & 1,1 \\ \hline \end{array} \qquad G_B: \quad \begin{array}{c|c|c|} & a_1 & a_0 \\ \hline a_1 & 1,1 & 0,0 \\ \hline a_0 & 0,0 & 0,0 \\ \hline \end{array} \qquad (1)$$

We shall refer to G_A and G_B as *state-games*. The probability with which Nature selects each state-game is not exactly known to the players.

2. The two players have a common prior $\mathbb{P}$ about the behavior of Nature.

3. Each player acquires private information about the choice of Nature. This information will be assumed to be binary. Thus, each player can be of two types, in the sense of Harsanyi ([5]). We shall say that player I is of type I_0 or of type I_1 according to whether he has seen, say, a Tail or a Head. Analogously for player II. For $k \in \{I, II\}$, we shall denote by $\mathbb{P}_{I_k}$ the conditional probability given the private information acquired by player I_k. We may think of $\mathbb{P}_{I_k}$ as the updated beliefs of I_k about the realized choice of Nature.

4. A binary public signal is shown to both players.

5. Each player chooses his action.

6. The state-game chosen by Nature is revealed and payoffs are collected accordingly.

Several criteria can be taken into account to select actions; among these, we consider the following:

CRITERION A: Each player chooses a_1 if and only if he thinks that G_B is more likely than G_A, conditionally on all the information available, private and public.

CRITERION B: The same as above, but not taking into account the public signal (namely, conditioning on private information only).

We may say that the phenomenon of *information rejection* occurs when both criteria lead to Nash equilibria and Criterion B is more favorable than Criterion A for at least one player. Recall that a Nash equilibrium is a strategy profile such that no player can profit from unilaterally deviating from his strategy in the profile.

Remark 1. Private information plays a crucial role. In fact, it is proved in [3] that in games with the structure described above, if the players have the same information they want as much information as possible. A more general result, relating positive value of information to uniqueness of Pareto optimal Nash equilibria is given in [2].

Remark 2. Although the game presented here is somehow artificial, it is in a sense the simplest possible example in which information refusal may occur. Private information (which is necessary, as we mentioned above) is binary, the action space is binary, public information is binary, the bimatrices of payoff have only one non-zero entry.

Remark 3. The rationale underlying the examples of information refusal which we are going to show can be phrased as follows. The games are coordination games, and the prior law is such that both players believe that G_B is more likely than G_A. It is known that one observation is not enough to reverse this opinion, but two observations may lead a player to believe that G_A is more likely. Thus, after one private observation is taken, the players may prefer to avoid an additional observation, in order not to run the risk of disrupting the initial coordination.

In order to characterize games as described above, we need to specify:

(a) The prior law $\mathbb{P}$ and the way it relates to the mechanism of choice by Nature
(b) The structure of private information and the way it helps to understand the unknown probability distribution of Nature on the two state-games.
(c) The structure of the public signal.

2.1 First Example: Hypothesis Testing

In this example we want to describe the situation when two (Bayesian) statisticians need to simultaneously test a simple hypothesis vs another simple hypothesis, and their payoff is positive iff both make the correct choice. We may think of G_A (resp.:G_B) as the payoffs when the true hypothesis is the null (resp.: the alternative).

We characterize the game along the lines sketched above.

(a) *Description of the prior.* The prior law $\mathbb{P}$ is a distribution on the parameter space $\Psi := \{\theta_0, \theta_1\}$, with $0 < \theta_0 < \theta_1 < 1$. The value θ_0 corresponds to the null hypothesis, and θ_1 to the alternative.
   We denote by π_0 the probability $\mathbb{P}$ that the state-game G_A is selected by Nature, i.e. that the null hypothesis holds true.
(b) *Structure of private information.* Let Θ be a Ψ-valued random variable such that $\Theta = \theta_0$ iff G_A is selected by Nature. Let also X_I, X_{II}, Y be random variables such that, conditionally on $\Theta = \theta$, they are i.i.d. Bernoulli with parameter θ, $\forall \theta \in \Psi$ (i.e. $\mathbb{P}(X_I = 1 | \Theta = \theta) = \theta$). It is common knowledge that the value of X_I is shown to player I only, and that the value of X_{II} is shown to player II only. Thus, each statistician has a private sample of size one from the population to be tested. Y is the public signal.
(c) *Structure of the public signal.* The value of Y is shown to both players. Thus, an additional sample of size one is observed by both statisticians.

A strategy profile in this game is a string of 8 actions: the first two are the actions taken by I_0 (i.e. Player I with private information $X_I = 0$) if $Y = 0$ and $Y = 1$, respectively, and so on. The following proposition shows that information refusal may occur.

Proposition 1. *Consider the game previously described. If the parameters* θ_0, θ_1 *and* π_0 *satisfy*

$$\left(\frac{1-\theta_1}{1-\theta_0}\right)^2 \max\left\{\frac{\theta_1}{1-\theta_0},1\right\} \leq \frac{\pi_0}{1-\pi_0} \leq \frac{1-\theta_1}{1-\theta_0} \min\left\{\frac{(\theta_1)^2}{\theta_0(1-\theta_0)},1\right\},$$

then:

1. *The following strategy profile*

$$\begin{array}{cccc} (I_0) & (I_1) & (II_0) & (II_1) \\ a_0 a_1 & a_1 a_1 & a_0 a_1 & a_1 a_1 \end{array}, \tag{2}$$

   is an equilibrium. Each action is the same that a single statistician would have taken if he were to maximize his expected utility based on all available information, namely, if he were to choose his action according to Criterion A.

2. *The following strategy profile*

$$\begin{array}{cccc} (I_0) & (I_1) & (II_0) & (II_1) \\ a_1 a_1 & a_1 a_1 & a_1 a_1 & a_1 a_1 \end{array}. \tag{3}$$

   is an equilibrium. Each action is the same that a single statistician would have taken if he were to maximize his expected utility based on private information only, namely, if he were to choose his action according to Criterion B.

3. *The payoff for* I_0 *if (2) is played is less than his payoff when (3) is played if and only if*

$$\frac{\pi_0}{1-\pi_0} \geq \left(\frac{1-\theta_1}{1-\theta_0}\right)^2 \frac{1}{(1-\theta_0)} \tag{4}$$

4. *The payoff for* I_1 *if (2) is played is always less than his payoff when (3) is played.*

Proof. First, we write the expressions for the payoffs:

- The expected payoff for I_0 if (2) is played is

$$A(\pi_0, \theta_0, \theta_1) := \mathbb{P}_{I_0}\left(\Theta = \theta_0, Y = 0, X_{II} = 0\right) + \mathbb{P}_{I_0}\left(\Theta = \theta_1, Y = 1\right). \tag{5}$$

- If player I_0 deviates from (2) and plays $a_1 a_1$ (other moves are clearly not advantageous) his expected payoff is

$$B(\pi_0, \theta_0, \theta_1) := \mathbb{P}_{I_0}\left(\Theta = \theta_1, Y = 0, X_{II} = 1\right) + \mathbb{P}_{I_0}\left(\Theta = \theta_1, Y = 1\right). \tag{6}$$

- The expected payoff for I_1 if (2) is played is

$$C(\pi_0, \theta_0, \theta_1) := \mathbb{P}_{I_1}\left(\Theta = \theta_1, Y = 0, X_{II} = 1\right) + \mathbb{P}_{I_1}\left(\Theta = \theta_1, Y = 1\right). \tag{7}$$

- If player I_1 deviates from (2) and plays $a_0 a_1$ (other moves are clearly not advantageous) his expected payoff is

$$D(\pi_0, \theta_0, \theta_1) := \mathbb{P}_{I_1} \left(\Theta = \theta_0, Y = 0, X_{II} = 0 \right) + \mathbb{P}_{I_1} \left(\Theta = \theta_1, Y = 1 \right). \quad (8)$$

- The expected payoff for I_0 if (3) is played is

$$E(\pi_0, \theta_0, \theta_1) := \mathbb{P}_{I_0} \left(\Theta = \theta_1 \right) \quad (9)$$

- The expected payoff for I_1 if (3) is played is

$$F(\pi_0, \theta_0, \theta_1) := \mathbb{P}_{I_1} \left(\Theta = \theta_1 \right) \quad (10)$$

It is clear that (3) is an equilibrium. In order to show that (2) is an equilibrium, we need to show that $A - B \geq 0$ and $C - D \geq 0$. In fact,

$$
\begin{aligned}
A(\pi_0, & \theta_0, \theta_1) - B(\pi_0, \theta_0, \theta_1) \\
&= \mathbb{P}_{I_0} \left(\Theta = \theta_0 Y = 0, X_{II} = 0 \right) - \mathbb{P}_{I_0} \left(\Theta = \theta_1 Y = 0, X_{II} = 1 \right) \\
&= \mathbb{P}_{I_0} \left(\Theta = \theta_0 \right) \mathbb{P}_{I_0} \left(Y = 0, X_{II} = 0 | \Theta = \theta_0 \right) \\
&\quad - \mathbb{P}_{I_0} \left(\Theta = \theta_1 \right) \mathbb{P}_{I_0} \left(Y = 0, X_{II} = 1 | \Theta = \theta_1 \right) \\
&= \frac{\pi_0 \left(1 - \theta_0 \right)}{\pi_0 \left(1 - \theta_0 \right) + \left(1 - \pi_0 \right) \left(1 - \theta_1 \right)} \left(1 - \theta_0 \right)^2 \\
&\quad - \frac{\left(1 - \pi_0 \right) \left(1 - \theta_1 \right)}{\pi_0 \left(1 - \theta_0 \right) + \left(1 - \pi_0 \right) \left(1 - \theta_1 \right)} \theta_1 (1 - \theta_1) \\
&\geq 0 \Leftrightarrow \frac{\pi_0}{1 - \pi_0} \geq \frac{\theta_1}{1 - \theta_0} \left(\frac{1 - \theta_1}{1 - \theta_0} \right)^2
\end{aligned}
$$

and

$$
\begin{aligned}
C(\pi_0, & \theta_0, \theta_1) - D(\pi_0, \theta_0, \theta_1) \\
&= \mathbb{P}_{I_1} \left(\Theta = \theta_1 Y = 0, X_{II} = 1 \right) - \mathbb{P}_{I_1} \left(\Theta = \theta_0 Y = 0, X_{II} = 0 \right) \\
&= \frac{\left(1 - \pi_0 \right) \theta_1}{\left(1 - \pi_0 \right) \theta_1 + \pi_0 \theta_0} \theta_1 (1 - \theta_1) - \frac{\pi_0 \theta_0}{\left(1 - \pi_0 \right) \theta_1 + \pi_0 \theta_0} \left(1 - \theta_0 \right)^2 \\
&\geq 0 \Leftrightarrow \frac{\pi_0}{1 - \pi_0} \leq \left(\frac{1 - \theta_1}{1 - \theta_0} \right) \frac{\theta_1^2}{\theta_0 (1 - \theta_0)}
\end{aligned}
$$

Next, we show that the actions in (2) (resp.: (3)) are those that a single statistician following Criterion A (resp.: Criterion B) would have chosen.

Preliminarily, we observe the following: if $\theta_0 < \theta_1$ and if $Z_1, Z_2, \ldots$ are i.i.d. conditionally on $\Theta = \theta$, for $\theta \in \{\theta_0, \theta_1\}$, with conditional distribution Bernoulli with parameter θ, then $\mathbb{P}(\Theta = \theta_1 | \sum Z_i = z)$ is increasing in z, as it is easy to check.

In view of these considerations, it is clear that we need only to show

$$\mathbb{P}_{I_0} \left(\Theta = \theta_1 | Y = 0 \right) < \frac{1}{2} < \mathbb{P}_{I_0} \left(\Theta = \theta_1 \right) \quad (11)$$

In fact,

$$\mathbb{P}_{I_0}\left(\Theta = \theta_1 | Y = 0\right) = \frac{\mathbb{P}\left(\Theta = \theta_1\right)\mathbb{P}\left(X_I = 0, Y = 0 | \Theta = \theta_1\right)}{\mathbb{P}\left(X_I = 0, Y = 0\right)}$$

$$= \frac{\left(1 - \pi_0\right)\left(1 - \theta_1\right)^2}{\left(1 - \pi_0\right)\left(1 - \theta_1\right)^2 + \pi_0\left(1 - \theta_0\right)^2}$$

$$< \frac{1}{2} \Leftrightarrow \frac{\pi_0}{1 - \pi_0} > \left(\frac{1 - \theta_1}{1 - \theta_0}\right)^2$$

Furthermore,

$$\mathbb{P}_{I_0}\left(\Theta = \theta_1\right) = \frac{\left(1 - \pi_0\right)\left(1 - \theta_1\right)}{\left(1 - \pi_0\right)\left(1 - \theta_1\right) + \pi_0\left(1 - \theta_0\right)} > \frac{1}{2} \Leftrightarrow \frac{\pi_0}{1 - \pi_0} < \frac{1 - \theta_1}{1 - \theta_0}.$$

Thus, (11) is proved, and the claim follows.

Now, we compare the payoffs of I_0 and I_1 in the two equilibria (2) and (3). It is easy to check that $C - F < 0$ for all values of the parameters. Hence, if given the choice, player I_1 would choose that the additional information Y not be revealed.

As far as I_0 is concerned, we must compare A and E:

$$A(\pi_0, \theta_0, \theta_1) - E(\pi_0, \theta_0, \theta_1)$$
$$= \mathbb{P}_{I_0}\left(\Theta = \theta_0 Y = 0, X_{II} = 0\right) - \mathbb{P}_{I_0}\left(\Theta = \theta_1 Y = 0\right)$$
$$= \frac{\pi_0(1 - \theta_0)}{\pi_0(1 - \theta_0) + (1 - \pi_0)(1 - \theta_1)}\left(1 - \theta_0\right)^2$$
$$- \frac{(1 - \pi_0)(1 - \theta_1)}{\pi_0(1 - \theta_0) + (1 - \pi_0)(1 - \theta_1)}\left(1 - \theta_1\right)$$
$$\geq 0 \Leftrightarrow \frac{\pi_0}{1 - \pi_0} \geq \left(\frac{1 - \theta_1}{1 - \theta_0}\right)^2 \frac{1}{(1 - \theta_0)}$$

Hence, only for high enough values of π_0 the payoff in the equilibrium emerging when Y is considered is higher. Thus, if given the choice, player I_0 would prefer that information be revealed for certain values of π_0 and would prefer that it be withheld for other values.

2.2   Second Example: Prediction

We want to describe here a situation in which two statisticians must simultaneously predict correctly a binary outcome in order to guarantee for themselves a positive reward.

The setup can be described as follows. Nature chooses repeatedly a state-game, each time with a probability Θ which is unknown to the players. They must predict which state-game Nature will choose next. Preliminary observations will help the players in assessing the value of Θ.

As we shall see below, we shall include in this example the possibility of partial signaling.

Here is a description of the game.

(a) *Description of the prior.* The prior law $\mathbb{P}$ determines the "a priori" distribution of Θ. We assume that this distribution is a Beta(α, β).

(b) *Structure of private information.* Let Y, X, X_I, X_{II} be exchangeable Bernoulli random variables, i.i.d. conditionally on Θ, with $\mathbb{P}(X = 1|\Theta = \theta) = \theta$. It is common knowledge that the value of X_I is shown to player I only, and that the value of X_{II} is shown to player II only. Y is the public signal (see below), and X represents the choice of Nature to be predicted: $X = 1$ if and only if the state-game G_B is selected.

(c) *Structure of the public signal.* A binary signal ξ_p is shown to both players $(0 \leq p \leq 1)$. It is such that, independently of the values of Y and of all random variables involved,

$$\mathbb{P}\left(\xi_p = Y\right) = p = 1 - \mathbb{P}\left(\xi_p = Z\right)$$

where Z is the outcome of a fair coin independent of X; thus, with probability p the r.v. ξ_p yields valuable information, namely Y, and with probability $1 - p$ it gives irrelevant information, namely the outcome of an independent coin toss ; for an example of such a variable, see the Remark below.

We may think of p as the clarity of the signal revealed. For each value of p we have a game, say G_p.

A strategy profile is described by a string of 8 actions. The first two are the actions taken by I_0 when $\xi_p = 0$ and $\xi_p = 1$, respectively, and so forth.

Remark 4. In order to describe the public signalling mechanism, consider first three independent Bernoulli random variables Y, W, Z such that

- Y, X_I, X_{II}, X are exchangeable;
- W is independent of X_I, X_{II}, X and $\mathbb{P}(W = 1) = p$;
- Z is independent of X_I, X_{II}, X and $\mathbb{P}(Z = 1) = \frac{1}{2}$.

The Bernoulli random variable ξ_p is described as follows: the coin W is tossed by a referee; if $W = 1$, then the value of Y is revealed, otherwise the fair coin Z is tossed and the result of the toss is revealed. Thus

$$\xi_p = \begin{cases} Y \text{ if } W = 1, \\ Z \text{ if } W = 0, \end{cases} \tag{12}$$

i.e.

$$\{\xi_p = k\} = \{W = 1, Y = k\} \cup \{W = 0, Z = k\}, \qquad k \in \{0, 1\}.$$

This mechanism is common knowledge, but the players don't know the outcome of W. They are only told the value of ξ_p (in addition to their private information). Observe that relevant information is given only when $W = 1$, which happens with probability p. If $p = 1$, then the players have an additional observation (exchangeable with X) before predicting X. If $p = 0$, then the additional observation Y is not available to the players.

Proposition 2. *Consider the games G_p as described above. If the parameters of the prior law of Θ satisfy the relation*

$$2 \leq \beta + 1 < \alpha < \beta + 2,$$

then:

1. *The following strategy profile*

$$\begin{array}{cccc} (\mathrm{I}_0) & (\mathrm{I}_1) & (\mathrm{II}_0) & (\mathrm{II}_1) \\ a_0 a_1 & a_1 a_1 & a_0 a_1 & a_1 a_1 \end{array}.$$

   (13)

   is an equilibrium. For

$$p > p_0 := \frac{1 - 2\frac{\beta+1}{\alpha+\beta+1}}{1 - 4\frac{\beta+1}{\alpha+\beta+1}\frac{\alpha}{\alpha+\beta+2}},$$

   each action is the same that a single statistician would have taken if he were to maximize his expected utility based on private information only, namely, if he were to choose his action according to Criterion A.

2. *The following strategy profile*

$$\begin{array}{cccc} (\mathrm{I}_0) & (\mathrm{I}_1) & (\mathrm{II}_0) & (\mathrm{II}_1) \\ a_1 a_1 & a_1 a_1 & a_1 a_1 & a_1 a_1 \end{array}.$$

   (14)

   is an equilibrium. Each action is the same that a single statistician would have taken if he were to maximize his expected utility based on private information only, namely, if he were to choose his action according to Criterion B. Furthermore, for $p < p_0$, each action is the same that a single statistician following Criterion A would have taken.

3. *For $p > p_0$, Criterion A is Pareto-dominated by Criterion B. Obviously, it leads to the same payoff for other values of p.*

Proof. First of all, since $\beta + 1 < \alpha$, we have

$$\mathbb{P}_{I_k}(X = 1) = \frac{\alpha + k}{\alpha + \beta + 1} > \frac{1}{2}, \qquad k = 0, 1.$$

Hence, it is obvious that Criterion B leads to (14), and this is clearly an equilibrium.

Next, we consider Criterion A. We have

$$\mathbb{P}_{I_0}(\xi_p = 0) = \mathbb{P}_{I_0}(\xi_p = Y, Y = 0) + \mathbb{P}_{I_0}(\xi_p = Z, Z = 0)$$

$$= p\frac{\beta + 1}{\alpha + \beta + 1} + (1 - p)\frac{1}{2},$$

and

$$\mathbb{P}_{I_0}(X = 0, \xi_p = 0) = p \frac{\beta+1}{\alpha+\beta+1}\frac{\beta+2}{\alpha+\beta+2} + (1-p)\frac{1}{2}\frac{\beta+1}{\alpha+\beta+1}.$$

Hence,

$$\mathbb{P}_{I_0}(X = 0|\xi_p = 0) = \frac{\frac{\beta+1}{\alpha+\beta+1}\left[p\frac{\beta+2}{\alpha+\beta+2} + (1-p)\frac{1}{2}\right]}{p\frac{\beta+1}{\alpha+\beta+1} + (1-p)\frac{1}{2}}$$

$$> \frac{1}{2} \Leftrightarrow p > \frac{1 - 2\frac{\beta+1}{\alpha+\beta+1}}{1 - 4\frac{\beta+1}{\alpha+\beta+1}\frac{\alpha}{\alpha+\beta+2}} = p_0.$$

Thus, the action of I_0 if he sees $\xi_p = 0$ is a_0. The other strategies in the profile can be established similarly.

Next, we show that the strategy profile (13) yields a Nash equilibrium of G_p, for every $p > p_0$. First, we write the expressions of the relevant payoffs:

- The payoff of I_0 in G_p if (13) is played is

$$\mathbb{P}_{I_0}(\xi_p = 0, X_{II} = 0, X = 0) + \mathbb{P}_{I_0}(\xi_p = 1, X = 1) \qquad (15)$$

$$= \frac{1-p}{2} \mathbb{P}_{I_0}(X_{II} = 0, X = 0) + p\, \mathbb{P}_{I_0}(Y = 0, Interactive X_{II} = 0, X = 0)$$

$$+ \mathbb{P}_{I_0}(\xi_p = 1, X = 1)$$

- The payoff of I_0 in G_p if he deviates and plays $a_1 a_1$ is

$$\mathbb{P}_{I_0}(\xi_p = 0, X_{II} = 1, X = 1) + \mathbb{P}_{I_0}(\xi_p = 1, X = 1) \qquad (16)$$

$$= \frac{1-p}{2} \mathbb{P}_{I_0}(X_{II} = 1, X = 1) + p\, \mathbb{P}_{I_0}(Y = 0, X_{II} = 1, X = 1)$$

$$+ \mathbb{P}_{I_0}(\xi_p = 1, X = 1)$$

We show now that the difference between (13) and (14), namely

$$\left(\frac{1-p}{2}\right)\left[\frac{\beta+1}{\alpha+\beta+1}\frac{\beta+2}{\alpha+\beta+2} - \frac{\alpha}{\alpha+\beta+1}\frac{\alpha+1}{\alpha+\beta+2}\right]$$

$$+ p\left[\frac{\beta+1}{\alpha+\beta+1}\left(\frac{\beta+2}{\alpha+\beta+2}\frac{\beta+3}{\alpha+\beta+3} - \frac{\alpha}{\alpha+\beta+2}\frac{\alpha+1}{\alpha+\beta+3}\right)\right]$$

is positive iff $p > p_0$. In fact, the above quantity is positive if and only if

$$p > \frac{\frac{\alpha}{\alpha+\beta+1}\frac{\alpha+1}{\alpha+\beta+2} - \frac{\beta+1}{\alpha+\beta+1}\frac{\beta+2}{\alpha+\beta+2}}{\frac{\alpha}{\alpha+\beta+1}\frac{\alpha+1}{\alpha+\beta+2} - \frac{\beta+1}{\alpha+\beta+1}\frac{\beta+2}{\alpha+\beta+2} + 2\frac{\beta+1}{\alpha+\beta+1}\left(\frac{\beta+2}{\alpha+\beta+2}\frac{\beta+3}{\alpha+\beta+3} - \frac{\alpha}{\alpha+\beta+2}\frac{\alpha+1}{\alpha+\beta+3}\right)}$$

Now, we see after some straightforward calculations that the right hand side equals p_0.

Next, we repeat the same arguments for player I_1. If (13) is played, his payoff is

$$\mathbb{P}_{I_1}(\xi_p = 1, X = 1) + \mathbb{P}_{I_1}(\xi_p = 0, X_{II} = 1, X = 1),$$

whereas if he deviates and plays $a_0 a_1$ his payoff becomes

$$\mathbb{P}_{I_1}(\xi_p = 1, X = 1) + \mathbb{P}_{I_1}(\xi_p = 0, X_{II} = 0, X = 0).$$

It is a simple matter to check that for every value of p there is no interest in deviating.

In order to prove the last claim of the Proposition, we first show that the payoff for I_0 if (14) is played , namely $\mathbb{P}_{I_0}(X = 1)$, is greater than (15). In fact, their difference yields

$$\mathbb{P}_{I_0}(\xi_p = 0, X = 1) - \mathbb{P}_{I_0}(\xi_p = 0, X_{II} = 0, X = 0)$$

$$= p \frac{\beta + 1}{\prod_{k=1}^{3}(\alpha + \beta + k)}[\alpha(\alpha + \beta + 3) - (\beta + 2)(\beta + 3)]$$

$$+ \frac{1 - p}{2\prod_{k=1}^{2}(\alpha + \beta + k)}[\alpha(\alpha + \beta + 2) - (\beta + 1)(\beta + 2)]$$

This difference is positive, since

$$\alpha(\alpha+\beta+3)-(\beta+2)(\beta+3) \geq (\beta+1)(2\beta+4)-(\beta+2)(\beta+3) = (\beta+2)(\beta-1) \geq 0,$$

and

$$\alpha(\alpha + \beta + 2) - (\beta + 1)(\beta + 2) \geq (\beta + 1)\alpha > 0.$$

It is even simpler to show that the payoff for I_1 is greater in the equilibrium (14) than in (15) In fact, if (14) is played, player I_1 collects a non-zero reward if and only if the event $\{X = 1\}$ occurs, whereas in (15) his reward is non-zero iff a proper subset of $\{X = 1\}$ occurs.

References

1. Avenhaus, R., von Stengel, B., Zamir, S. (1995). Inspection games. Preprint.
2. Bassan, B., Gossner, O., Scarsini, M., Zamir, S. (2000). On the value of information in interactive decision systems. Preprint.
3. Bassan, B., Scarsini, M., Zamir, S. (1998). Uniqueness of Pareto optima, coordination games and positive value of information. Preprint.
4. Bassan, B., Scarsini, M., Zamir, S. (2001). Role of information in the interaction of two statisticians: some game theoretic results. In *Recent Developments in Operational Research*, Agarwal, M.L., Sen, K. (Eds.), 33–43, Narosa, New Delhi estimation,
5. Harsanyi, J.C. (1967/68). Games with incomplete information played by 'Bayesian' players, Parts I, II, and III. Managem. Sci. **14**, 159–182, 320–334, 486–502

Probabilistic Modelling: An Historical and Philosophical Digression

Antonio Forcina

Department of Statistics, University of Perugia, Perugia, Italy

Abstract. This paper is about the conflict between the modern formal treatment of statistical inference and the role of subjectivity, inventiveness and personal involvement which, I claim, should be allowed in any non trivial applied probabilistic modelling. I concentrate, intentionally, on the limitations of the formal treatment and try to overemphasize the qualitative, informal judgments involved in applied inference. Overdispersion and Item Response models are used as an illustration.

1 Away from Formal Statistical Inference

By *Formal Inference* I roughly mean the cultural paradigm underlying almost all theoretical formulations about drawing statistical inferences. The usual story goes like this: we have observations about a random variable whose distribution depends on unknown parameters. I have no objection to this for the situations where the probability distribution under consideration is produced by sampling from a finite population or by random assignment of treatments to units, where modelling is indeed trivial. In most other cases, as I will try to show below, to describe the randomness within which the data may be embedded is like inventing a story which, however, has to be consistent in its own terms.

Formal Inference deals with idealized situations; as such the relevance of its results for those who work with applications may be similar to the relevance of grammar for good writing. The most common form in which the tools from mathematical statistics may be used is that, if we pretend that a set of very restrictive assumptions hold true, then we can assess how much the data support a single hypothesis of interest. This is fine as long as we do not forget that it is based on a good deal of fiction.

1.1 A Brief Historical Digression

I wish I had the competence to trace the origins of this paradigm since the early formulations of the method of least squares until when Fisher in [3] phrased it in a way which is essentially the same that we use today. Roughly speaking, in the formulation of astronomical problems whose solutions led to the method of least squares (see for instance Stigler [11], Chap. 1), there were rather well defined physical quantities which could not be observed without error and they were treated as unknowns in a system of linear equations, with each equation corresponding to an observation.

It may be of interest to notice that the problem of estimating, for a given earthquake, the coordinates of the hypocenter and the time of the event, which is a genuine statistical problem, is formulated in modern seismology ([6], p. 221-224) in exactly the same way. Let τ be the time of the event and t_i the time of arrival of the P-wave at the ith station, λ be the coordinates of the hypocenter and y_i those of the ith station. Under the assumption that the waves travel at constant speed μ, one can write down a system of equations of the form

$$t_i = \tau + \frac{\sqrt{(y_i - \lambda)'(y_i - \lambda)}}{\mu}, \quad i = 1,\ldots,n$$

which expresses the relation between the time of departure of the wave, the distance, the time of arrival and the speed. Here τ and λ are what we call *parameters* but to the seismologist they are just unknowns in a over determined system of equations.

When parameters are no longer well defined physical quantities, we enter into the realm of fantasy. In itself this is not a bad thing as long as we keep in mind that, though parameters should provide the answer to relevant questions of interest in the problem at hand, often they exist only in the model which is defined through them. For a study about the origin of the word 'parameter' see the contribution of [10] to the discussion of Leonard. My impression is that Fisher and his contemporaries while using the word *parameter* were aware of real applications of which they had direct personal experience. Today instead we have many theoretical statisticians who are very familiar with the formal properties of the theory but have very little contact and interest for the problems to which the theory could be applied.

1.2 Keynes or Coming Back to the Real Thing

An important part of Keynes' book *A Treatise on Probability*, which first appeared in 1920, is devoted to a criticism of Statistical Inference. Though the discussion is based on works published before 1920 and thus Keynes does not seem to be aware of Fisher's contributions, the main points in his criticism of Statistical Inference, contained in Part V of the book, seem to me to be extremely relevant to a modern statistician. In a way, Keynes' criticism brings us back to the real issues underlying any inference, including those based on statistical methods. However, my assessment of the relevance of Keynes in this respect is probably not shared by many statisticians: see for example [12] for additional references and a very critical evaluation.

According to Keynes, the statistician is faced with all the difficulties which are inherent to inductive reasoning: a good deal of the knowledge which must be taken into account is of a vague nature and incapable of numerical treatment ([5], p. 328). What makes the life of the statistician easier is that, part of the information which is to be taken into account, is available in a convenient and manageable form, summarized into statistics. This situation, however, may be misleading if one believe that statistics is all that one needs to consider. Keynes in [5], p. 391, wrote:

No one supposes that a good induction can be arrived at merely by counting cases. The business of strengthening the argument chiefly consists in determining whether the alleged association is stable when the accompanying conditions are varied. This process of improving the Analogy, as I have called it in Part III, is, both logically and practically, at the essence of the argument.

In other words, the strength of an inductive conclusion must increase with the diversity and complexity of the experimental conditions under which, for example, a given treatment is effective. Often these issues will be discussed when planning the data collection or designing an experiment and involve a lot of qualitative assessments about which set of circumstances might affect the result or association of interest. Now, according to the formal treatment of the subject in statistics, it seems that, by simply increasing the number of observations, we can also increase the strength of an empirical finding. Suppose for instance that we want to compare two fertilizers; if all the observations are taken in the same area, in a given period and under similar conditions, we clearly have very little support for extending what has been observed, even if the number of trees used in the experiment is very large. Moreover, usually, once the data have been collected, only very simple information will be available concerning the design and the methods of observation. The problem is even more serious when, as it is customary in scientific papers, one tries to reanalyze popular data sets which are, usually, almost completely abstracted from the context in which they were produced.

A closely connected problem is that the strength of an inference must also depend on how wide and general is the statement we are willing to assert. The fact is that there is no way for expressing such features within Formal Statistical Inference. Consider again the example of a designed experiment to examine the effectiveness of different fertilizers; our conclusions might be valid for predicting the results for the same area, in the same period and possibly under very similar atmospheric conditions. But with these restrictions there is almost no inference but simply a description of what did happen in a very specific context.

Though these issues are usually ignored in the statistical literature, I think that they are somehow related to the notion of *overfitting*. I do not know of any generally accepted definition of *overfitting*, but I have seen the notion used occasionally by referees to mean roughly that, if the model fits very well, it is likely that some dirty trick has been used. A more serious assessment of overfitting should look for indirect signs that the model is likely to be making statements about *accidental facts specific of the observed data*. To take an extreme example, in a linear model context, by inspecting a large number of possible contrasts, it should not be difficult to find a small subset of these contrasts whose estimates are highly significant and a model stating that the remaining linearly independent contrasts are 0 will probably fit very well. Thus, the essential quality of *overfitting* is not that the model fits well, but that it is making assertions about randomness and this should be revealed by the fact that what the model asserts looks complicated and uninteresting.

2   Where Do Models Come From?

The title of this section echoes a similar title in a paper by Lehmann [7] that complained that Fisher had apparently very little to say but "As regards problems of specification, these are entirely a matter for the practical statistician". In the following I will try to discuss the issue from a very *personal* point of view and highlight certain connections between Statistics and the Arts.

In many instances probabilistic models may be seen as the outcome of a dialog between two parties which I will call *the statistician* and *the scientist* with the understanding that more than one person may be involved on both sides. *The statistician* will usually know very little about the specific field until *the scientist* comes along with the data or a research project. The quality of the dialog that will be established will affect the quality of the resulting work to an extent which is comparable to that due to the technical abilities of each party in their own fields. To begin with, each side will be speaking in a different *language*. So, it is likely that *the scientist*, when asked to describe the applied context, will describe how things should behave according to some preliminary model which is not acknowledged as such. The starting point of the scientific investigation or *the scientist*'s expectations may be inconsistent or rely on assumptions which, once translated into a probabilistic framework, may turn out to be vague or meaningless.

On the other hand, *the statistician* will often be tempted to pay little attention to details that do not fit easily into the probabilistic framework which seemed initially appropriate. Indeed it is often the case that something which initially may seem irrelevant or simply a nuisance, at a closer look may reveal interesting features. For instance, in a study of the population dynamics of a species of crustacea, an excess of variability, initially accommodated into a model of overdispersion, led later to examine more closely the sampling process. It turned out that *the scientists* had provided an over simplified description of the sampling process and that what had been explained by a strange feature of the spatial distribution of the crustacea was instead due to the fact that only a small proportion of the material collected from the lake was actually examined. So, while *the scientist* may forget to disclose important piece of information, especially if this could bring discredit upon her/his work, *the statistician* is likely to be reassuringly inaccurate when asked to translate the probabilistic assumptions into *the scientist* own language.

The way of questioning the other party and of being alerted against the potential misunderstandings described above is something that can only be learned by watching other people doing it and is an instance of what Polanyi [9] calls *personal skills*. Two other important notions which are particularly relevant here and were also developed by Polanyi are those of *personal involvement* and *intellectual passions*. Even in a simple context, if we do not take anything for granted, the actual class of possible models would be very wide and even if we were able to explore all of them, we would hardly find our way out of the Labyrinth. If some structure will eventually emerge, this will be to some extent an invention rooted in the reality as well as into our own curiosity and obsessions.

2.1   Is Statistics an Art?

This is again a quote from [7]:

> Is applied statistics, and particularly model building, an art with each
> new case having to be treated from scratch (although even artistic
> endeavors require technique that can be systematized and learned),
> completely on its own or does theory has a contribution to make to
> this process?

Consider how, within our community, we decide which are the relevant fields of research and how we assess the merits of single research projects: certainly not on the basis of formal or objective criteria, rather more often we are guided by our emotional response telling us that something goes into the right direction. So this is another instance of Polanyi's *intellectual passions* which however may be trained and it is mainly through personal contacts that we statisticians come to share some common feelings about the style of probabilistic modelling. It is easy to see that these attitudes have much in common with the poetics of artistic movements.

The main issue here is: how much space there is for subjectivity and inventiveness in applied probabilistic modelling and how tight is the constraint that the model has to fit the data after all. If the model has to contain a generalization, it must aim to catch only those features of the data which are of interest. So, for instance, if there is no reason to suspect that the order with which cases are observed makes any difference, this information will not even be taken into account and even if we noticed an apparent systematic effect, this will probably be attributed to chance. In other cases it is the model itself that says how its adequacy should be assessed; a particularly interesting instance is that of binomial models with overdispersion: it is as if these models were born with an alibi for not fitting as expected.

2.2 Models or Fantasies? An Example from Item Response Models

To make my discussion more specific I will try to apply it to item response theory. This is essentially a flexible set of probabilistic models aimed to represent the behavior of n subjects, selected at random from a given population, who submit to an examination made up of J dichotomous items (questions). In the latent class version of these models, one imagines that the population is made up of C different latent classes of individuals which are homogeneous with respect to the abilities needed to answer the items correctly. Let x denote a $J \times 1$ vector of possible responses, that is a string of 0's and 1's that will be called response configuration; clearly there are 2^J possible configurations and we may denote with p_c the vector whose elements are the probabilities of providing all possible response configuration (ordered, say, lexicographically) conditionally on latent class c; this vector describes completely the behavior of subjects in latent class c.

Most item response models assume that the events of giving a correct answer to different items are independent, conditionally on a given latent class and that latent classes may be ordered in a unique way from the worst to the best (with respect to the probability of answering correctly any given item). A substantial simplification is achieved by the Rasch model which, in the context of finite mixtures (see [8]), assumes that the difference between the conditional logits for any pair of items is constant across latent classes and depends only on the differential difficulties of the items.

2.3 Over or Underdispersion?

Clearly, latent classes are just fiction and any inference will have to be based on the so-called manifest distribution, that is marginally on latent classes; the corresponding data are contained in the vector of observed frequencies y giving the number of subjects classified according to the response configuration they provide, irrespective of the latent class to which they belong. Now the question is: what is the probability distribution of y or, at least, what is its variance matrix? Clearly, if this distribution was multinomial, we would have

$$Var(y) = n[diag(p) - pp'] = n\Omega(p),$$

where p is an appropriate vector of marginal probabilities having the same structure as p_c. The matter is not so trivial as it may appear and in fact [2] in the context of capture recapture data, devoted an Appendix to show that the distribution is almost multinomial though dispersion is less than multinomial. A simple proof of this last statement is as follows. Let y_c be the frequency distribution for the n_c subjects belonging to latent class c. If we let $p = \sum n_c p_c / n$, the variance of y conditionally on $n = (n_1 \ldots n_C)$ and $P = (p_1 \ldots p_C)$, is simply the sum of the (conditional) multinomial variances and may be transformed by adding and subtracting npp' so that

$$\sum n_c[diag(p_c) - p_c p_c'] = n\Omega(p) - \sum n_c(p_c - p)(p_c - p)'$$

and the claim follows from the fact that the first component is a multinomial variance and the second component is a positive definite matrix.

What is not clear, however, is why we should condition to so many quantities which are unobservable and exist only in our fiction. A different model arise if we assume, for instance, that the n_c (the number of subjects sampled from each latent class) are fixed but the p_c are random with $E(p_c) = p$ and $Var(p_c) = V$. Computations are straightforward but a lot more tedious in this case and are omitted, however the result is well know and has been used, for instance, by Brown and Payne ([1]) to model overdispersion in the context of electoral data. The variance of y may be written as $n\{\Omega(p) + [\sum(n_c^2/n - 1)]V\}$ and the amount of overdispersion depends on the second component.

A somehow surprising result arise if we assume that the p_c are constant while the number of subjects sampled from each latent class follow a multinomial distribution: in this case the variance of the manifest distribution is exactly multinomial. Essentially this is so because the additional dispersion induced by the sampling variation of the n_c compensate exactly the underdispersion of the initial model, which was equivalent to a mixture of conditional multinomial distributions. More precisely, if we let the n_c have a multinomial distribution with expectation $n\pi_c$ and π denote the vector of probabilities with elements π_c, then the variance of $E(y) = Pn$ may be written as $nP\Omega(\pi)P'$ and this is equal to $n\sum\pi_c(p_c - p)(p_c - p)'$.

All of these models, and many others, were described by Gini ([4]), pp 151–154, in his study of the distribution of sexes in human births. Each model is formulated without ambiguity by specifying in detail the random procedure that can generate the data. For instance, for the second model above he assumes that we first select with replacement a given number n of balls from a box containing balls numbered from 1 to C and obtain the sample sizes n_c; then, for each c from 1 to C, select n_c balls from a box having a composition based on p_c. However, in Gini's applications, over or under dispersion are testable assumptions because he was considering the distribution of families with a given number of children according to the number of sons and daughters and not according to all possible configurations of sex in the children ordered by age. In other words, in Gini's context the manifest distribution may be compared against the binomial distribution. Instead, in the context of item response (or capture-recapture) data, with the single table y the issue is undecidable.

References

1. Brown, P.J., Payne, C.D. (1986). Aggregate data, ecological regression and voting transition. J. Am. Stat. Assoc. **81**, 452–460

2. Darroch, J.N., Fienberg, S.E., Glonek, G.F.V., Junker, B.W. (1993). A three sample multiple-recapture approach to census population estimation with heterogeneous catchability. J. Am. Stat. Assoc. **88**, 1137–1148
3. Fisher, R.A. (1935). The logic of inductive inference. J. Roy. Stat. Soc. A. **98**, 39–54
4. Gini, C. (1908). *Il Sesso (dal Punto di Vista Statistico)*. Sandroni, Napoli
5. Keynes, J.M. (1948). *A Treatise on Probability*. Macmillan and Co., London
6. Lay, T., Wallace, T.C. (1995). *Modern Global Seismology*. Academic Press, San Diego
7. Lehmann, E.L. (1990). Model specification: the views of Fisher and Neyman, and later developments. Stat. Sci. **5**, 160–168
8. Lindsay, B., Clogg, C., Grego, J. (1991). Semiparametric estimation in the Rasch model and related exponential response models, including a simple latent class model for item analysis. J. Am. Stat. Assoc. **86**, 96–107
9. Polanyi, M. (1954). *Personal Knowledge*. MacMillan, London
10. Stigler, S.S. (1976). Contribution to Discussion of "On Re-reading R. A. Fisher," by L.J. Savage. Ann. of Stat. 4, 498–500
11. Stigler, S.M. (1986). *The History of Statistics*. Harward University Press, Cambridge, Massachussets
12. Stigler, S.M. (1999). *Statistics on the Table*. Harward University Press, Cambridge, Massachussets

A Bayesian View on Sampling the 2 × 2 Table

Seymour Geisser

School of Statistics, University of Minnesota, Minneapolis, USA

Abstract. We study exact and approximate inferential procedures for the 2 × 2 table from both the frequentist and Bayesian mode mediated by Likelihood Principles. In particular, for a variety of sampling rules, inferential procedures for a Bayesian approach are the same while differences ensue for various exact and some approximate conditional frequentist methods. In fact, for certain sensible sampling rules, no exact conditional frequentist procedure is available. In a hypothetical situation where it is assumed that the sampling rule that led to the table was unknown, suggestions are made to handle this case, that indicate the general superiority and versatility of the Bayesian approach.

1 Introduction

Inference from the 2 × 2 table has been a source of interest and dispute for almost as long as statistics has been a modern discipline. Initially there was the Fisher-Pearson dispute over the proper degrees of freedom for the chi-squared test of independence (or equality of two population proportions). Later came the Fisher-Neyman dispute, see Barnard [1], [2], on an appropriate exact small sample test. It would appear that Fisher [12] had always claimed that whether one was sampling from a multinomial or two binomials the appropriate exact test of significance for independence in the former and equality of the population proportions in the latter was to condition on the marginals. The sampled values and true probabilities are displayed in Table 1. In this paper we discuss the various sampling approaches that would give rise to such a table and methods of analysis for particular parameters involved in the table.

Table 1.

	A	$\overline{A}$	
B	p_{11}	p_{12}	n_1
	r_1	$n_1 - r_1$	
$\overline{B}$	p_{21}	p_{22}	n_2
	r_2	$n_2 - r_2$	
	r	$n - r$	n

If we are dealing with this classical 2×2 table then the random values within the table have the multinomial probability function

$$f_n = f(r_1, r_2, n_1 - r_1, n_2 - r_2) = \frac{n!\, p_{11}^{r_1}\, p_{12}^{n_1-r_1}\, p_{21}^{r_2} p_{22}^{n_2-r_2}}{r_1!\, r_2!\, (n_1 - r_1)!\, (n_2 - r_2)!} \tag{1}$$

subject to the four arguments summing to n and $\sum_{i,j} p_{ij} = 1$, with prescribed sampling size n.

If we reparametrize to

$$p_1 = p_{11}/(p_{11} + p_{12}), p_2 = p_{21}/(p_{21} + p_{22}), \gamma = p_{11} + p_{12}$$

then

$$\begin{aligned}
f_n &= f(r_1, r_2 | n_1, n_2) f(n_1 | n) \\
&= \binom{n_1}{r_1} p_1^{r_1} (1 - p_1)^{n_1 - r_1} \binom{n_2}{r_2} p_2^{r_2} (1 - p_2)^{n_2 - r_2} \binom{n}{n_1} \gamma^{n_1} (1 - \gamma)^{n_2}.
\end{aligned} \tag{2}$$

Now inference about a function $g(p_1, p_2)$, will be the same from f_n as from the first two factors or the rhs of (2) by invoking an extension of the usual Likelihood Principle, LP (see Appendix), that was proposed by Barnard et al. [7] and formulated explicitly by Barnard and Sprott [6], which owed a great deal to Fisherian ideas. Potential restrictions and extensions and variations were discussed by Barnard [3], [4], Barnard and Sprott [6], Basu [8] and Berger and Wolpert [9]. Here invoking one extension termed ELP (see Appendix) is equivalent to conditioning on say n_1 so that n_2 is also given since n was already fixed, which yields the first two terms of (2).

If we now condition one of the other marginal sums, say $r = r_1 + r_2$, then $n - r$ is also fixed and we have, as Fisher indicated, conditioned on all of the marginals. This yields

$$f(r_1 | r, n_1, n_2) = \binom{n_1}{r_1} \binom{n_2}{r - r_1} \Psi^{r_1} \Bigg/ \sum_a^b \binom{n_1}{j} \binom{n_2}{r - j} \Psi^j \tag{3}$$

$b = \min(r, n_1)$ and $a = \max(r - n_2, 0)$ $\Psi = p_1(1 - p_2)/p_2(1 - p_1)$.

Equal tailed conditional confidence limits can be obtained numerically by obtaining solutions $\Psi = \Psi_2$ the upper limit and $\Psi = \Psi_1$ the lower limit from

$$\sum_{r_1=0}^{c} f(r_1 | r, n_1, n_2) = \frac{\alpha}{2}, \quad \sum_{r_1=c}^{n_1} f(r_1 | r, n_1, n_2) = \frac{\alpha}{2} \tag{4}$$

respectively, so that with at least confidence $1 - \alpha$, we obtain an interval (Ψ_1, Ψ_2). Approximate large sample conditional confidence intervals for Ψ using (3) were obtained by Cornfield [11] as well.

He showed that in the limit, using Sterling's formula

$$-2\log \frac{f(r_1|r,n_1,n_2)}{f(\tilde{r}_1|r_1 n_1, n_2)} = (r_1 - \tilde{r}_1)^2/\tau^2$$

where $\tilde{r}_1$ is the mode of the distribution of r_1 and

$$\tau^2 = \left[\frac{1}{\tilde{r}_1} + \frac{1}{n_1 - \tilde{r}_1} + \frac{1}{r - \tilde{r}_1} + \frac{1}{n_2 - r - \tilde{r}_1}\right]^{-1}.$$

By approximating a sum by an integral he showed that in the limit

$$f(\tilde{r}_1|r_1 n_1, n_2) = \frac{1}{\sqrt{2\pi}}\tau^{-1},$$

thus demonstrating that the limiting distribution of (3) is $N(\tilde{r}_1, \tau^2)$. Hence, accommodating a correction for continuity, he set

$$(\tilde{r}_1 - r_1 - \frac{1}{2})^2\tau^{-2} = \chi_\alpha^2 \tag{5}$$

$$(\hat{r}_1 - r_1 + \frac{1}{2})^2\tau^{-2} = \chi_\alpha^2 \tag{6}$$

where χ_α^2 is the upper α percent point of the chi-square distribution with one degree of freedom. The largest real root r_u of the equation (5) in $\tilde{r}_1$ and the smallest real root r_s of the equation (6) form the $1 - \alpha$ confidence limits on $\tilde{r}_1$.

He then showed that for sufficiently large samples

$$\Psi \doteq \frac{\tilde{r}_1(n_2 - r + \tilde{r}_1)}{(n_1 - \tilde{r}_1)(r - \tilde{r}_1)}. \tag{7}$$

Therefore Ψ is a monotonic function of $\tilde{r}_1$ and approximate conditional confidence limits on Ψ, (Ψ_1, Ψ_2) are obtained by substituting r_s and r_u in (7).

At this point we note that the likelihood of the first two factors of (2) which are independent binomials, is

$$f_{IB} = \binom{n_1}{r_1}\binom{n_2}{r_2}\Psi^{r_1}(1 - p_2)^{n-r}p_2^r \Big/ \left(1 + \tfrac{p_2}{1-p_2}\Psi\right)^{n_1} \tag{8}$$

Hence the ELP would be contravened for estimation about Ψ because as a function of Ψ (3) and (8) are not proportional. Here a Bayesian analysis would in general yield different inferences for Ψ depending on whether (3) or (8) was used with the same prior density for Ψ.

Fisher's exact significance test for independence in the 2×2 table for the equality of p_1 and p_2 from the independent binomials, where under the null hypothesis $\Psi = 1$, reduces to the use of the standard hypergeometric probability function

$$f(r_1|r, n, n_1, n_2) = \frac{\binom{n_1}{r_1}\binom{n_2}{r-r_1}}{\binom{n}{r}} \qquad (9)$$

so there is a disconnect in terms of consistently applying the ELP to simple significance testing and estimation with regard to Fisherian inference. This was already noted by Barnard [5] who nevertheless supported the exact test because he claimed that little information was lost in conditioning on the marginals. However it also turned out that this conditional test is basically a similar test if, within the Neyman-Pearson corpus of hypothesis testing randomization is included (see Tocher [16]).

2   Sampling Issues

There are many other ways of sampling that would lead to a 2×2 table. For example, we can allow n to be random (negative multinomial sampling) and condition on any one of the marginals or tabular entries. Suppose then for n random we stop sampling until a fixed value of n_1 is achieved. We then find

$$f_{n_1} = f(r_1, r_2, n|n_1) = f(r_1, r_2|n_1, n)f(n|n_1) \qquad (10)$$
$$= \binom{n_1}{r_1}p_1^{r_1}(1-p_1)^{n_1-r_1}\binom{n_2}{r_2}p_2^{r_2}(1-p_2)^{n_2-r_2}\binom{n-1}{n_1-1}\gamma^{n_1}(1-\gamma)^{n_2}$$

However the likelihood for p_1 and p_2 is still the same although the overall sampling distribution is obviously different than the usual multinomial. Hence inference about functions of p_1 and p_2, according to the ELP, is still the same as when we assumed multinomial sampling.

Negative multinomial sampling can also occur if one sampled n until a fixed r is achieved. In this case we get

$$f_r = f(r_1, n_1, n_2|r) \qquad (11)$$
$$= f(n_1, n_2|r_1, r_2)f(r_1|r)$$
$$= \binom{n_1-1}{r_1-1}p_1^{r_1}(1-p_1)^{n_1-r_1}\binom{n_2-1}{r_2-1}p_2^{r_2}(1-p_2)^{n_2-r_2}\binom{r}{r_1}\alpha^{r_1}(1-\alpha)^{r_2}$$

where $\alpha = p_{11}/(p_{11} + p_{21})$.

Although the likelihood for p_1 and p_2 arises from two independent negative binomials it is the same as in the positive multinomial and the independent binomials case. However, a frequentist can condition on $n_1 + n_2$ yielding a sampling probability function

$$f(n_1|r_1, r_2, n) = \frac{\binom{n_1-1}{r_1-1}\binom{n-n_1-1}{r_2-1}\theta^{n_1}}{\sum_{r_1}^{n-r_2}\binom{j-1}{r_1-1}\binom{n-j-1}{r_2-1}\theta^j}. \qquad (12)$$

where $\theta = \frac{1-p_1}{1-p_2}$, i.e. the ratio of the failure probabilities. Conditional confidence intervals can be obtained for θ along the lines of Cornfield's treatment of Ψ. (By reversing the definition of failure and success one can parametrize to the ratio of success probabilities). Here the parametricization differs from (3) and the likelihood from (2) which is also the likelihood for independent negative binomials. Again the ELP is not sustained. But a simple frequentist significance test for $\theta = 1$ is equivalent to $p_1 = p_2$ and results in the standard negative hypergeometric probability function

$$f(n_1|r_1,r_2,n) = \frac{\binom{n_1-1}{r_1-1}\binom{n-n_1-1}{r_2-1}}{\binom{n-1}{r_1+r_2-1}}. \tag{13}$$

Such a solution has been implicitly suggested by Lehmann [15]. Whether or not this adheres to the Fisherian significance testing outlook, of course, cannot readily be determined—but it would appear so. Further, it can be shown to be basically a similar test in Neyman-Pearson framework, if randomization is included, as was Fisher's exact test (5).

Conditional confidence levels can be obtained numerically from solutions $\theta = \theta_2$, the upper limit, and $\theta = \theta_1$, the lower limit from

$$\sum_{n_1=a}^{c} f(n_1|r_1,r_2,n) = \frac{\alpha}{2}, \quad \sum_{n_1=c}^{b} f(n_1|r_1,r_2,n) = \frac{\alpha}{2}. \tag{14}$$

Large sample conditional confidence limits can be obtained in a manner similar to Cornfield's treatment described in section 1. We note that for large samples it is easy to ascertain that

$$\theta \doteq \frac{(n-\tilde{n}_1)(\tilde{n}_1-r_1)}{\tilde{n}_1(n-\tilde{n}_1-r_2)}, \tag{15}$$

where $\tilde{n}_1$ is the mode of n_1. Further, the limiting distribution of n_1 is normal with mean $\tilde{n}_1$ and variance

$$\tau^2 = \left[\frac{1}{\tilde{n}_1} + \frac{1}{\tilde{n}_1-r_1} + \frac{1}{n-\tilde{n}_1} + \frac{1}{n-\tilde{n}_1-r_2}\right]^{-1}. \tag{16}$$

Hence denote n_u as the largest real root of the equation in $\tilde{n}_1$ of

$$\frac{(\tilde{n}_1-n_1-\frac{1}{2})^2}{\tau^2} = \chi_\alpha^2 \tag{17}$$

and n_s as the smallest real root of

$$\frac{(\tilde{n}_1-n_1+\frac{1}{2})^2}{\tau^2} = \chi_\alpha^2 \tag{18}$$

where n_1 is observed and χ_α^2 defined as the upper α per cent point of the chi-square distribution with one degree of freedom. Then solutions for upper and lower $1-\alpha$ limits on θ are obtained by substituting the limits on $\tilde{n}_1$, $(\tilde{n}_s, \tilde{n}_u)$ from (17) and (18) in (15) to obtain (θ_1, θ_2).

3   The Mixed Sampling Case

Another negative multinomial sampling approach stops when r_1 attains a given value. Here

$$
\begin{aligned}
f_{r_1} &= f(r_2, n_1, n | r_1) = f(r_2, n | n_1, r_1) f(n_1 | r_1) \\
&= f(n_1 | r_1) f(r_2 | n, n_1, r_1) f(n | n_1, r_1) \\
&= \binom{n_1 - 1}{r_1 - 1} p_1^{r_1} (1 - p_1)^{n_1 - r_1} \binom{n_2}{r_2} p_2^{r_2} (1 - p_2)^{n_2 - r_2} \binom{n - 1}{n_1 - 1} \gamma^{n_1} (1 - \gamma)^{n_2}
\end{aligned}
\tag{19}
$$

Again, the likelihood for p_1 and p_2 is preserved for Bayesian inference but here we now encounter a difficulty for conditonal frequentist inference regarding p_1 and p_2. What does one condition on to obtain an exact significance test on $p_1 = p_2$, or a similar test in Neyman-Pearson setup? Of course this would also occur when we start with one sample that is Binomial, say a control, and the other negative Binomial, for say a new treatment where one would like to stop the latter trial after a given number of failures.

In this situation, while Bayesian inference is not altered, exact frequentist inference appears to be stymied, whether for testing or estimation. So we have not only a disconnect between testing and estimation for conditional Fisherian frequentist inference if the ELP is to be obeyed, but more disconcerting a stonewall for the mixed case.

However, the Bayesian approach also suffers from a slight disconnect for these testing cases as the usual continuous prior for p_1 and p_2 for estimation is no longer appropriate for testing $p_1 = p_2$ because the posterior probability for the null hypothesis would be zero. The usual method to circumvent this is to put a lump of probability on the null hypothesis (see Jeffreys [14]). So even though this Bayesian approach is not completely seamless, a slight modification also turns the testing procedure into one of model selection, see also Geisser [13], Bernardo and Smith [10].

4   What Equal Likelihoods Entail

Suppose we are only told that in a series of independent and identically distributed binary trials there were r successes and $n - r$ failures, and the sampling was conducted in one of three ways:

1. The number of trials was fixed at n.
2. Sampling was stopped at the $r - th$ success.
3. Sampling was stopped when $n - r$ failures were obtained.

Now while the three sampling probabilities differ they all have the same likelihood

$$
L = p^r (1 - p)^{n - r}.
$$

The probability of r successes and $n-r$ failures under these sampling methods are

$$f_n = \binom{n}{r} L, \ \ f_r = \binom{n-1}{r-1} L, \ \ f_{n-r} = \binom{n-1}{n-r-1} L$$

where f_a denotes the probability where a is fixed for sampling.

Suppose we are to infer which one of these 3 sampling rules was used in the absence of any other information. This would be of interest since a frequentist analysis would depend on the sampling rule.

If prior probabilities of the rules are assumed and denoted as p_n, p_r, and p_{n-r} where $p_n + p_r + p_{n-r} = 1$, then the probability of the rule that gave rise to the table is

$$P(S_n) \propto \frac{n}{r(n-r)} p_n, \ \ P(S_r) \propto \frac{1}{n-r} p_r, \ \ P(S_{n-r}) \propto \frac{1}{r} p_{n-r}$$

respectively, where S_a denotes the sampling rule with a fixed. If we assume $p_n = p_r = p_{n-r} = 1/3$, a "possible" ignorance assumption, then clearly

$$P(S_n) = \frac{1}{2}, \ \ P(S_r) = \frac{r}{2n} \ \text{ and } \ P(S_{n-r}) = \frac{n-r}{2n}$$

so that

$$P(S_n) \geq \max(P(S_r), \ P(S_{n-r}))$$

with equality holding for either $r = n$ or 0 (this result informs us which of the two negative binomial sampling rules could not have occurred).

On the other hand one might intuit that one really should not discriminate between the rules based on the information given so that a posteriori $P(S_n) = P(S_r) = P(S_{n-r}) = 1/3$, for $r \neq 0$ or n, then $p_n \propto r(n-r), p_r \propto n(n-r)$ and $p_{n-r} \propto rn$. One could easily object to such a set of prior probabilities (aside from their peculiarity) because of their dependence on the values of the data.

A third view is that statistical inference is incalculable in such a situation except subjectively since in this hypothetical situation no other information is presumed. However a subjective view, that could accept the largest probability for sampling rule S_n, is that this guarantees that the experiment will terminate whereas the other two plans cannot guarantee that the experiment will not be indefinitely long. Since S_n does have the largest likelihood and subsequent largest probability under prior ignorance this appears to be a reasonable inference. Note also that if $r > n - r$ that sampling rule S_r denominates S_{n-r} and vice-versa if $n - r > r$.

Next we examine these issues for the 2×2 table. Here we list the various ways one can sample in constructing a 2×2 table such that one of the 9 values that are seen is fixed in the sense that when that value appears sampling ceases. For 7 out of the 9 cases the entire likelihood is the same where

$$L = \gamma^{n_1} (1-\gamma)^{n_2} \prod_{i-1}^{2} p_i^{r_i} (1-p_i)^{n_i - r_i} = L(\gamma) L(p_1, p_2)$$

with sampling probability

$$f_n = \binom{n_1}{r_1}\binom{n_2}{r_2}\binom{n}{n_1}L,$$

$$f_{n_1} = \binom{n_1}{r_1}\binom{n_2}{r_2}\binom{n-1}{n_1-1}L, \quad f_{n_2} = \binom{n_1}{r_1}\binom{n_2}{r_2}\binom{n-1}{n_2-1}L,$$

$$f_{r_1} = \binom{n_1-1}{r_1-1}\binom{n_2}{r_2}\binom{n-1}{n_1-1}L, \quad f_{r_2} = \binom{n_1}{r_1}\binom{n_2-1}{r_2-1}\binom{n-1}{n_2-1}L,$$

$$f_{n_1-r_1} = \binom{n_1-1}{n_1-r_1-1}\binom{n_2}{r_2}\binom{n-1}{n_1-1}L,$$

$$f_{n_2-r_2} = \binom{n_1}{r_1}\binom{n_2-1}{n_2-r_2-1}\binom{n-1}{n_2-1}L.$$

The other two whose total likelihoods differ from the above are still equivalent to the above for inference on (p_1, p_2) by the virtue of the ELP. But for inference on the sampling rules one requires the non-extended LP, since ELP does not apply. They are

$$f_r = \binom{n_1-1}{r_1-1}\binom{n_2-1}{r_2-1}\binom{r}{r_1}L(p_1,p_2)\alpha^{r_1}(1-\alpha)^{r_2}$$

$$f_{n-r} = \binom{n_1-1}{n_1-r_1-1}\binom{n_2-1}{n_2-r_2-1}\binom{n-r}{n-r_1}L(p_1,p_2)\beta^{n_1-r_1}(1-\beta)^{n_2-r_2}$$

where

$$\alpha = p_{11}/(p_{11}+p_{21}), \quad \beta = p_{12}/(p_{12}+p_{22}).$$

Restricting our attention to the initial 7 sampling rules whose total likelihoods are equal we consider the same issue as previously. Can we infer, upon being presented only with the entries of a 2×2 table, which of the 7 sampling rules were used to generate the table, assuming only one of those 7 rules was actually used.

If we assume all of the 7 were equally likely to be used then the probability of each of the sampling rules is

$$P(S_n) = \frac{1}{3}, \ P(S_{n_1}) = \frac{n_1}{3n}, \ P(S_{n_2}) = \frac{n_2}{3n}, \ P(S_{r_1}) = \frac{r_1}{3n}, \ P(S_{r_2}) = \frac{r_2}{3n},$$

$$P(S_{n_1-r_1}) = \frac{n_1-r_1}{3n}, \ P(S_{n_2-r_2}) = \frac{n_2-r_2}{3n}$$

where S_a represents the sampling rule until a is achieved. Clearly S_n dominates all the others except in the unusual case where there are at least two zeros among the tabulated values involved with table 1, then at least one other rule will also have a 1/3 probability. Other than this unusual case, the second most probable is S_{n_i} where $n_i = \max(n_1, n_2)$.

After that the probability of S_a depends on the size of a – the larger a the larger the posterior probability of S_a. On the other hand we could force the posterior probabilities to be all equal – but as before there are arguments that appear to mitigate against such a view. Of course the third possibility that a reasonable inference without further information, (such as the knowledge of the sequence of trials) and some subjective information, is unavailable. It is also clear that f_{r_2}, $f_{n_1-r_1}$ and $f_{n_2-r_2}$ are in the same category as f_{r_1} in that an exact conditional frequentist approach is unavailable, while all the others can be handled by the methods detailed in sections 1 and 2.

5 Remarks

In summary it is to be noted that for the Bayesian who is inferring about $g(p_1, p_2)$ it really does not matter which of the 9 sampling rules generated the data as long as $g(p_1, p_2)$ is independent of γ or α or β. However, a frequentist interested in exact or in approximating exact inference would be in difficulty when ignorant of the sampling rule. Although this is a situation that admittedly does not occur with great frequency (assuming contact with the generator of the table), but when it does a Bayesian approach can be used to decide on the sampling rule, i.e. select the rule with largest probability or base it on the resulting mixture. However, willingness to use a Bayesian approach for deciding on the sampling rule should also favor the use of the Bayesian approach on $g(p_1, p_2)$ and in fact avoid deciding on the sampling rule. When the prior distribution of p_1 and p_2 is assumed independent of the remaining parameter of the reparamatrization then the ELP is completely in accord with the Bayesian approach. This reinforces the view that there are cogent theoretical and practical reasons for treating the 2×2 table in a Bayesian mode.

Appendix

Likelihood Principle (LP)

Preliminaries

The model for experiment $\mathcal{E}$ consists of a sample space S and a parametric space Θ and a family of probability functions $f : S \times \Theta \longrightarrow R^+$ such that for all $\theta \in \Theta$

$$\int_S f d\mu = 1.$$

LP

For two such experiments modeled as $\mathcal{E} = \{S, \mu, \Theta, f\}$ $\mathcal{E}' = \{S', \mu', \Theta, f'\}$, and for realization $X \in S$ and $X' \in S'$, if

$$f(x|\theta) = g(x, x')f'(x'|\theta) \text{ for } g > 0$$

for all θ and the choice of $\mathcal{E}$ or $\mathcal{E}'$ is uniformative with regard to θ then inference or information $\text{Inf}(\mathcal{E}, X) = \text{Inf}(\mathcal{E}', X')$ concerning θ.

Note this implies that all of the statistical evidence provided by the data is conveyed by the likelihood function. There is an often useful extension namely:

ELP

When $\theta = (p, \gamma)$ and

$$f(x|p, \gamma) = g(x, x', \gamma)f'(x'|p)$$

it is plausible to extend LP to

$$\text{Inf}(\mathcal{E}, X) = \text{Inf}(\mathcal{E}', X') \text{ concerning } p,$$

assuming that p and γ are unrelated.

References

1. Barnard, G.A. (1945) A New Test for 2×2 Tables. Nature. **156**, 177.
2. Barnard, G.A. (1949) Statistical Inference. Journal of the Royal Statistical Society, B. Vol. II, 115–139.
3. Barnard, G.A. (1964) Lecture notes (delivered at the National Institute of Health).
4. Barnard, G.A. (1971) Scientific Inferences and Day-to-Day Decisions. Foundations of Statistical Inference, ed. Godambe, V.P. and Sprott, D.A. Holt, Rhinehart and Winston. 163–176.
5. Barnard, G.A. (1979) In Contradiction to J. Berkson's Dispraise: Conditional Tests can be More Efficient. Journal of Statistical Planning Inference. **3**, 181–187.
6. Barnard, G.A. and Sprott, D.A. (1971) A Note on Basu's Examples of Anomalous Ancillary Statistics. *Foundations of Statistical Inference*, ed. Godambe, V.P. and Sprott, D.A. Holt, Rhinehart and Winston. 289–230.
7. Barnard, G.A. et al. (1962) Likelihood Inference and Time Series. Journal of the Royal Statistical Society, A. **125**, 321–372.
8. Basu, D. (1977) On the Elimination of Nuisance Parameters. Journal of the American Statistical Association. **358**, 355–366.

9. Berger, J.O. and Wolpert, R.L. (1984) *The Likelihood Principle*. IMS Lecture Notes, ed. Gupta, S.S. Vol. 6.

10. Bernardo, T.M. and Smith, A.F.M. (1994) *Bayesian Theory*. John Wiley and Sons, Chichester.

11. Cornfield, J. (1956) A Statistical Problem Arising from Retrospective Studies. Proceedings of the third Berkeley Symposium on Mathematical Statistics and Probability. University of California Press, Berkeley. 135–148.

12. Fisher, R.A. (1956) *Statistical Methods and Scientific Inference*. Oliver and Boyd, Edinburgh.

13. Geisser, S. (1993) *Predictive Inference: An Introduction*. Chapman and Hall, London.

14. Jeffreys, H. (1961) *Theory of Probability*, 3rd Edition. Oxford University Press, London.

15. Lehmann, E. L. (1991) *Testing Statistical Hypothesis*, 2nd edition. Wadsworth and Brooks/Cole, Pacific Grove, California.

16. Tocher, K.D. (1950) Extension of the Neyman-Pearson Theory of Test to Discontinuous Variates. Biometrika. **37**, 130–144.

Bayesian Designs for Binomial Experiments

Athanassios Katsis[1] and Blaza Toman[2]

[1] Department of Statistics and Actuarial Science, University of the Aegean,
   Samos, Greece
[2] Statistical Engineering Division, National Institute of Standards and
   Technology, Gaithersburg, USA

Abstract. Calculating the size of the sample required for an experiment is of paramount importance in statistical theory. We describe a new methodology for calculating the optimal sample size when a hypothesis test between two or more binomial proportions takes place. The posterior risk is computed and should not exceed a pre-specified level. A second constraint examines the likelihood of the unknown data not satisfying the bound on the risk.

1 Introduction

The heuristic argument for deriving the optimal sample design is straightforward but powerful. Initially, we place a constraint in the desired *precision* . Then, this precision is expressed mathematically in terms of the sample size n. When conducting a hypothesis test between two binomial proportions, the normal approximation to the binomial distribution is utilized and the power of the test involves n.

Several criteria have so far been proposed for Bayesian sample size estimation in the binomial setting. In [1], [2] and [3] a tolerance region is proposed where the parameters of a multinomial distribution will be contained with a certain probability. In [8] the sample sizes are obtained by imposing precision conditions on the posterior variance and Bayes risk whereas in [5] the pre-posterior marginal distribution of the data is employed to derive the sample size for intervals with either fixed length or fixed probability of coverage.

Generally, in a Bayesian development, there is no interest in Type I or Type II error probabilities. Precision is measured through posterior accuracy as in [4] in the context of one way ANOVA. Using the "$0 - 1$" loss function, the posterior risk is: $\min\{Pr(H_o|y), Pr(H_1|y)\}$. For this to be small, the following upper bound condition is imposed:

$$\min\{Pr(H_0|y), Pr(H_1|y)\} \leq \varepsilon \qquad (1)$$

Furthermore, since sample size calculation occurs before any data collection an additional condition is necessary to ensure that the set $K = T^c$ of all the data not satisfying (1) has a small probability of occurrence. i.e.,

$$Pr(y \notin T) = Pr(y \in K) < \delta \qquad (2)$$

where δ is a small constant and the probability in (2) is calculated based on the marginal distribution of y.

2   Formulation of the Problem

Let Y_1 and Y_2 be independent binomial random variables with parameters n_1, p_1 and n_2, p_2 respectively. We wish to derive the optimal total sample size, n, for performing the test of hypotheses:

$$H_0 : p_1 = p_2 = p \qquad vs \qquad H_1 : p_1 \neq p_2$$

where $n = n_1 + n_2$. We shall consider the case when p is a fixed number, and the case when p is not specified. The allocation ratio will be maintained as $n_1 = n_2 = \frac{n}{2}$.

The prior probability of H_0 is π_0 and of H_1 is $\pi_1 = 1 - \pi_0$. The prior information on the two proportions is expressed in terms of two independent Beta distributions, i.e., $p_i \sim \text{Beta}(\alpha_i, \beta_i)$. As stated before, the posterior risk should be bounded i.e., $\min\{P(H_0|y), P(H_1|y)\} \leq \varepsilon$ and the data $y = (y_1, y_2)$ must satisfy this bound with a high probability, i.e. $Pr(y \notin T) < \delta$ where T is the set of all y satisfying (1). The latter condition ensures that it is unlikely that data contradicting (1) will appear.

2.1   Case 1: p Is Known

Let us first examine the case that the proportion p has a known value. Under the null hypothesis the posterior density of p_1 and p_2 is:

$$g(p, p|y) = \frac{\left(\prod_{i=1}^2 \binom{n/2}{y_i}\right) p^{y_1+y_2} q^{n-y_1-y_2} \pi_0}{m(y)}$$

Under the alternative hypothesis, the joint posterior density of p_1 and p_2 is given by:

$$g(p_1, p_2|y) = \frac{\pi_1 \prod_{i=1}^2 \binom{n/2}{y_i} p_i^{\alpha_i+y_i-1} q_i^{n/2+\beta_i-y_i-1} \frac{\Gamma(\alpha_i+\beta_i)}{\Gamma(\alpha_i)\Gamma(\beta_i)}}{m(y)}$$

where $q_i = 1 - p_i$ and

$$m(y) = \left[\prod_{i=1}^2 \binom{n/2}{y_i} p^{y_i} q^{n/2-y_i}\right] \pi_0$$
$$+ \prod_{i=1}^2 \left[\int_0^1 \binom{n/2}{y_i} p_i^{\alpha_i+y_i-1} q_i^{n/2+\beta_i-y_i-1} \frac{\Gamma(\alpha_i+\beta_i)}{\Gamma(\alpha_i)\Gamma(\beta_i)} dp_i\right] \pi_1$$

The posterior odds are then given by

$$\frac{P(H_0|y)}{P(H_1|y)} = B\frac{\pi_0}{\pi_1}$$

where the Bayes factor B is given by

$$B = \frac{p^{y_1+y_2}q^{n-y_1-y_2}\prod_{i=1}^{2}\Gamma(\alpha_i+n/2+\beta_i)\Gamma(\alpha_i)\Gamma(\beta_i)}{\prod_{i=1}^{2}\Gamma(\alpha_i+y_i)\Gamma(n/2+\beta_i-y_i)\Gamma(\alpha_i+\beta_i)}$$

Next, it is necessary to find the set $K = T^c$. It is clear that

$$P(H_0|y) = \frac{(\pi_0/\pi_1)B}{1+(\pi_0/\pi_1)B}$$

$$P(H_1|y) = \frac{1}{1+(\pi_0/\pi_1)B}$$

Hence, it is straightforward to show that

$$K = \left\{ y : \frac{\varepsilon}{1-\varepsilon}\frac{\pi_1}{\pi_0} < B < \frac{1-\varepsilon}{\varepsilon}\frac{\pi_1}{\pi_0} \right\} \tag{3}$$

The following general result by [6] establishes that the marginal probability of the set K converges to zero as the sample size increases to infinity:

Theorem 1. *Let n denote the sample size. As $n \to \infty$, the Bayes factor B converges to 0 or ∞.*

The exact sample size is determined by solving for n in the following equation:

$$P\left\{ \frac{\varepsilon}{1-\varepsilon}\frac{\pi_1}{\pi_0} < B < \frac{1-\varepsilon}{\varepsilon}\frac{\pi_1}{\pi_0} \right\} = \delta \tag{4}$$

By considering α_i's and β_i's to be integers and using Sterling's approximation, we can write (4) approximately as follows: $P(W_1 < f(y_1, y_2) < W_2) = \delta$ where W_1 and W_2 are constants with respect to the data the data y and

$$f(y_1, y_2) = \prod_{i=1}^{2} \frac{p^{y_i}}{q^{y_i}(\alpha_i + y_i - 1)^{(\alpha_i+y_i-0.5)}(n/2+\beta_i-y_i-1)^{(n/2+\beta_i-y_i-0.5)}}$$

By taking the natural logarithms of W_1, W_2, and $f(y_1, y_2)$ we obtain the following expression for the sample size:

$$P(W_1' < g(y_1, y_2) < W_2') = \delta \tag{5}$$

where $W_1' = \ln W_1$, $W_2' = \ln W_2$ and $g(y_1, y_2) = \ln f(y_1, y_2)$. We will use Taylor's theorem to approximate (5). By evaluating all the partial derivatives at $y_i = \frac{n_i}{2} = \frac{n}{4}$, the Taylor approximation of $g(y_1, y_2)$ yields:

$$g(y_1, y_2) = g\left(\frac{n}{4}, \frac{n}{4}\right) + \sum_{i=1}^{2}\left(y_i - \frac{n}{4}\right)\frac{\partial g}{\partial y_i}\left(\frac{n}{4}, \frac{n}{4}\right)$$

$$+ \frac{1}{2!}\left\{\sum_{i=1}^{2}\left(y_i - \frac{n}{4}\right)^2\frac{\partial^2 g}{\partial y_i^2}\left(\frac{n}{4}, \frac{n}{4}\right)\right\}$$

where

$$\frac{\partial g}{\partial y_i}\left(\frac{n}{4},\frac{n}{4}\right) = \ln\frac{p}{q} - \frac{\alpha_i + \frac{n}{4} - 0.5}{\alpha_i + \frac{n}{4} - 1} + \frac{\beta_i - \frac{n}{4} - 0.5}{\beta_i - \frac{n}{4} - 1}$$

$$- \ln\left(\alpha_i + \frac{n}{4} - 1\right) + \ln\left(\beta_i - \frac{n}{4} - 1\right)$$

$$\frac{\partial^2 g}{\partial y_i^2}\left(\frac{n}{4},\frac{n}{4}\right) = -\frac{1}{\alpha_i + \frac{n}{4} - 1} + \frac{1}{2(\alpha_i + \frac{n}{4} - 1)^2}$$

$$- \frac{1}{\beta_i - \frac{n}{4} - 1} + \frac{1}{2(\beta_i - \frac{n}{4} - 1)^2}$$

Note that $\frac{\partial^2 g}{\partial y_1 \partial y_2} = 0$. Now, if we define the following quantities:

$c_1 = g\left(\frac{n}{4},\frac{n}{4}\right) - \frac{n}{4}\sum_{i=1}^{2}\frac{\partial g}{\partial y_i} + \frac{n^2}{32}\sum_{i=1}^{2}\frac{\partial^2 g}{\partial y_i^2}$, $c_2 = \frac{\partial g}{\partial y_1} - \frac{n}{4}\frac{\partial^2 g}{\partial y_1^2}$, $c_3 = \frac{\partial g}{\partial y_2} - \frac{n}{4}\frac{\partial^2 g}{\partial y_2^2}$, $c_4 = \frac{1}{2}\frac{\partial^2 g}{\partial y_1^2}$, $c_5 = \frac{1}{2}\frac{\partial^2 g}{\partial y_2^2}$, we have that (5) can be written as

$$P(W_1'' < y_1 c_2 + y_2 c_3 + y_1^2 c_4 + y_2^2 c_5 < W_2'') = \delta \qquad (6)$$

where $W_1'' = W_1' - c_1$, and $W_2'' = W_2' - c_1$.

The quantities c_4 and c_5 are almost always negative. For negative c_4 and c_5 the expression $h(y_1, y_2) = y_1 c_2 + y_2 c_3 + y_1^2 c_4 + y_2^2 c_5$ represents a quadratic surface in y_1 and y_2 with a maximum at $M = -\frac{c_2^2}{4c_4} - \frac{c_3^2}{4c_5}$. A solution to this inequality exists in the following cases:

- For $W_1'' < M < W_2''$ the inequality is satisfied for the points y_1 and y_2 that fall in the interior of the ellipse defined by $h(y_1, y_2) = W_1''$.
- For $W_1'' < W_2'' < M$ the inequality is satisfied for the points y_1 and y_2 that fall in the interior of the ellipse defined by $h(y_1, y_2) = W_1''$ and the exterior of the ellipse defined by $h(y_1, y_2) = W_2''$

From the above analysis, we conclude the following.

$$P(K) = P(y_1, y_2 \in \text{interior } h(y_1, y_2) = W_1''|W_1'' < M < W_2'') +$$
$$P(y_1, y_2 \in \text{interior } h(y_1, y_2) = W_1'' \text{ and}$$
$$y_1, y_2 \notin \text{interior } h(y_1, y_2) = W_2''|W_1'' < W_2'' < M).$$

Optimal sample sizes are now approximately found by solving: $P(K) = \delta$.

2.2   Case 2: $H_0 : p_1 = p_2$, p Unknown

In this situation, we are interested in testing $H_0 : p_1 = p_2$, the common value p is not of particular interest. This can be modelled by letting a priori $p \sim$ Beta(α, β). The prior information on the proportions p_i's is still expressed with Beta distributions, i.e., $p_i \sim$ Beta(α_i, β_i). Following the same methodology as before, the posterior odds are given by

$$B_1 = \frac{P(H_0|y)}{P(H_1|y)} = B\frac{\pi_0}{\pi_1}$$

where the Bayes factor B is

$$B = \frac{\prod_i [\Gamma(\alpha_i)\Gamma(\beta_i)\Gamma(\alpha_i + n/2 + \beta_i)]\Gamma(\alpha + \beta)\Gamma(\sum y_i + \alpha)\Gamma(n + \beta - \sum y_i)}{\prod_i [\Gamma(\alpha_i + \beta_i)\Gamma(\alpha_i + y_i)\Gamma(n/2 + \beta_i - y_i)]\Gamma(\alpha)\Gamma(\beta)\Gamma(n + \alpha + \beta)}$$

Using the same reasoning as before we obtain the set K:

$$K = \left\{ y : \frac{\varepsilon}{1 - \varepsilon}\frac{\pi_1}{\pi_0} < B < \frac{1 - \varepsilon}{\varepsilon}\frac{\pi_1}{\pi_0} \right\} \tag{7}$$

Again, as a direct consequence of the result by [6] the marginal probability of the set K converges to zero as the sample size increases to infinity.

Using the Taylor expansion the sample size equation is transformed to

$$P(V_1' < y_1 d_1 + y_2 d_2 + y_1^2 d_3 + y_2^2 d_4 + y_1 y_2 d_5 < V_2') = \delta \tag{8}$$

where V_1 and V_2 are constants with respect to the data and

$$\begin{aligned}
h(y_1, y_2) = {} & \left(\sum_{i=1}^2 y_i + \alpha - 0.5 \right) \ln \left(\sum_{i=1}^2 y_i + \alpha - 1 \right) \\
& + \left(n + \beta - \sum_{i=1}^2 y_i - 0.5 \right) \ln \left(n + \beta - \sum_{i=1}^2 y_i - 1 \right) \\
& - \sum_{i=1}^2 \{ (\alpha_1 + y_i - 0.5) \ln(\alpha_i + y_i - 1) \\
& \qquad - (n/2 + \beta_i - y_i - 0.5) \ln(n/2 + \beta_i - y_i - 1) \}
\end{aligned}$$

and $V_i' = \ln V_i - h(\frac{n}{4}, \frac{n}{4}) - \frac{n}{4}(\frac{\partial h}{\partial y_1} + \frac{\partial h}{\partial y_2}) + \frac{n^4}{32}(\frac{\partial^2 h}{\partial y_1^2} + \frac{\partial^2 h}{\partial y_2^2}) + \frac{n^2}{16}\frac{\partial^2 h}{\partial y_1 \partial y_2}$, $d_1 = \frac{\partial h}{\partial y_1} - \frac{n}{4}\frac{\partial h}{\partial y_1^2}$, $d_2 = \frac{\partial h}{\partial y_2} - \frac{n}{4}\frac{\partial h}{\partial y_2^2}$, $d_3 = \frac{1}{2}\frac{\partial h}{\partial y_1^2}$, $d_4 = \frac{1}{2}\frac{\partial h}{\partial y_2^2}$ and $d_5 = \frac{\partial^2 h}{\partial y_1 \partial y_2}$.

All the partial derivatives are evaluated at $\frac{n}{4}$. In this case the sign of the coefficients of the quadratic terms cannot be easily determined. The probability in (8) is transformed to:

$$\begin{aligned}
& P\left(d_3 y_1^2 + d_1 y_1 + d_5 y_1 y_2 + d_4 y_2^2 + d_2 y_2 - V_1' > 0 \text{ and} \right. \\
& \qquad \left. d_3 y_1^2 + d_1 y_1 + d_5 y_1 y_2 + d_4 y_2^2 + d_2 y_2 - V_2' < 0 \right) \\
& = \sum_{y_2} P(d_3 y_1^2 + d_1 y_1 + d_5 y_1 y_2 + d_4 y_2^2 + d_2 y_2 - V_1' > 0 \text{ and} \\
& \qquad d_3 y_1^2 + d_1 y_1 + d_5 y_1 y_2 + d_4 y_2^2 + d_2 y_2 - V_2' < 0 | y_2) f(y_2) \tag{9}
\end{aligned}$$

The roots of the quadratic forms with respect to y_1 and y_2 are denoted by $y_{1,+}$, $y_{1,-}$ and $y_{2,+}$, and $y_{2,-}$ respectively. We use the following notation: $X_+ = \max(y_{1,+}, y_{1,-})$, $X_- = \min(y_{1,+}, y_{1,-})$, $Z_+ = \max(y_{2,+}, y_{2,-})$, and

$Z_- = \min(y_{2,+}, y_{2,-})$. Examining the signs of the quadratic forms we obtain the following:

$$
\begin{aligned}
P(K) = &\sum_{\{y_2:d_3>0 \wedge L_1<0 \wedge L_2>0\}} P(Z_- < y_1 < Z_+|y_2)f(y_2) \\
+ &\sum_{\{y_2:d_3<0 \wedge L_1>0 \wedge L_2<0\}} P(X_- < y_1 < X_+|y_2)f(y_2) \\
+ &\sum_{\{y_2:d_3>0 \wedge L_1>0 \wedge L_2>0\}} P(Z_- < y_1 < X_-|y_2)f(y_2) \\
+ &\sum_{\{y_2:d_3>0 \wedge L_1>0 \wedge L_2>0\}} P(X_+ < y_1 < Z_+|y_2)f(y_2) \\
+ &\sum_{\{y_2:d_3<0 \wedge L_1>0 \wedge L_2>0\}} P(X_- < y_1 < Z_-|y_2)f(y_2) \\
+ &\sum_{\{y_2:d_3<0 \wedge L_1>0 \wedge L_2>0\}} P(Z_+ < y_1 < X_+|y_2)f(y_2) = \delta
\end{aligned}
$$

Hence, the optimal sample sizes are obtained by solving for n in $P(K) = \delta$.

3   Results

In this section, the method previously presented is illustrated by some specific examples. The algorithm is based on the argument used in section 2.2. The optimal sample size is found when $P(K) = \delta$. The program is written in SAS.

Table 1 shows the optimal sample sizes for the case presented in section 2.1. The parameters p, α_i and β_i are given in the table, while we have set $\pi_0 = 0.5$, $\varepsilon = 0.1$ and $\delta = 0.3$.

The specific cases presented in Table 1 highlight the features of the prior specification which have the greatest effect on the optimal sample size. Firstly, for a given precision δ, we need a bigger sample size to detect any significant differences when a priori the two proportions are expected to be closer to each other than when they are expected to be more distant. As an example, we consider the following two cases: In the first case, the prior mean proportions are 0.2 and 0.3 whereas in the second case they are set at 0.2 and 0.5. The fixed value p is 0.2 in both cases. As expected, the former situation requires three times the sample size of the latter (120 to 40) since the distance between the prior means increases considerably (0.1 to 0.3).

Another feature of the prior specification that determines the sample size is the distance between the prior mean proportions and the fixed value p. As this distance becomes smaller, it is increasingly more difficult to distinguish between the two proportions and p, hence the sample size increases. The following example illustrates this fact: Assume that in both cases the two proportions have the same prior mean 0.2, but in one case p is 0.2 and in the other is 0.5. The former case yields a sample size of $n = 152$ whereas the latter one needs only 20 observations.

There is an interplay between sample sizes and the prior variances of the p_i's. Consider, the case when $p = 0.5$, $p_1 = 0.5$ and $p_2 = 0.6$ and also when $p = 0.2$, $p_1 = 0.2$ and $p_2 = 0.3$. The sample sizes are somewhat different (138 to 120). The differences between the prior mean proportions and p are the same in both cases. Hence, the difference in sample sizes can be attributed to the differences of the prior variances. In the former case the prior variances are 0.036 and 0.022 for p_1 and p_2 respectively while in the latter case the variances are 0.027 and 0.019. In this situation a larger sample is required to detect a difference when there is more uncertainty about the prior proportions.

Another observation can be made. Let's examine the two cases when $p = 0.2$, while the prior means are either both 0.3 or both 0.1. The sample sizes are somewhat different (116 to 134). When $E(p_i) = 0.3$, the prior variance is 0.019, while in the case of $E(p_i) = 0.1$, it is a much smaller 0.004. Thus, in the second case although we are more certain about the prior information, more sampling is required to detect a difference. This is explained by the small values of the prior mean proportions ($E(p_i) = 0.1$ in the latter case compared to $E(p_i) = 0.3$ initially). Therefore, the prior probabilities of "success" in the second case are very small and we need a bigger sample size to detect a significant difference than in the first case. Finally, note that when the prior variances are not different, and the distances between the prior mean proportions and p are the same, we do not observe any difference in the resulting sample size ($n = 40$ for $p = 0.2$, $E(p_1) = 0.2$, $E(p_2) = 0.5$, as well as for $p = 0.5$, $E(p_1) = 0.5$, $E(p_2) = 0.8$). Thus the effect of size of the prior mean proportions appears to be the weakest of the four factors.

The results when the null hypothesis is $H_o : p_1 = p_2$ can be summarized as follows. The general behavior of the optimal sample size is again governed by the distance of the prior means $E(p_1)$ and $E(p_2)$. The further apart these are, the smaller the optimal sample size is.

Comparing the above results to the ones obtained when testing among three or more binomial populations (see [7]) , we note that the overall sample sizes tend to be smaller as the number of populations increase. This is a rather intuitive conclusion, since in order to establish H_1 in the case of, say, three populations a difference needs to be detected in just one of the three pairs among p_1, p_2 and p_3, whereas in the case of two distributions the comparison was strictly between p_1 and p_2.

In conclusion, the method presented in this article provides a fully Bayesian solution to the problem of sample size determination for hypothesis testing in the case of two binomial proportions.

Table 1. Sample size values for fixed p and $\varepsilon = 0.10$, $\delta = 0.30$

p	$E(p_1)$	$Var(p_1)$	$E(p_2)$	$Var(p_2)$	α_1,β_1	α_2,β_2	n
0.2	0.2	0.027	0.2	0.027	1,2	1,4	152
0.5	0.5	0.036	0.5	0.036	3,3	3,3	152
0.8	0.8	0.027	0.8	0.027	4,1	4,1	152
0.2	0.2	0.027	0.3	0.019	1,4	3,7	120
0.5	0.5	0.036	0.6	0.022	3,3	4,6	138
0.8	0.8	0.027	0.9	0.004	4,1	18,2	130
0.2	0.2	0.027	0.1	0.004	1,4	2,18	152
0.5	0.5	0.036	0.4	0.022	3,3	4,6	146
0.8	0.8	0.027	0.7	0.019	4,1	7,3	128
0.2	0.3	0.019	0.3	0.019	3,7	3,7	116
0.5	0.6	0.022	0.6	0.022	6,4	6,4	140
0.8	0.9	0.004	0.9	0.004	18,2	18,2	130
0.2	0.1	0.004	0.1	0.004	2,18	2,18	134
0.5	0.4	0.022	0.4	0.022	4,6	4,6	138
0.8	0.7	0.019	0.7	0.019	7,3	7,3	116
0.2	0.2	0.027	0.5	0.036	1,4	3,3	40
0.5	0.5	0.036	0.8	0.027	3,3	4,1	40
0.5	0.5	0.036	0.2	0.027	3,3	1,4	40
0.8	0.8	0.027	0.5	0.036	4,1	3,3	42
0.2	0.5	0.036	0.5	0.036	3,3	3,3	20
0.5	0.8	0.027	0.8	0.027	4,1	4,1	20
0.5	0.2	0.027	0.2	0.027	1,4	1,4	20
0.8	0.5	0.036	0.5	0.036	3,3	3,3	20

References

1. Adcock, C.J. (1987). A Bayesian approach to calculating sample sizes for multinomial sampling. Statistician **36**, 155–159
2. Adcock, C.J. (1988). A Bayesian approach to calculating sample sizes. Statistician **37**, 433–439
3. Adcock, C.J. (1992). Bayesian approaches to the determination of sample sizes for binomial and multinomial sampling–some commments on the paper by Pham-Gia and Turkkan. Statistician **41**, 399–404

4. DasGupta, A., Vidakovic, B. (1997). Sample size problems in ANOVA: Bayesian point of view. J. Stat. Plan. Infer. **65**, 335–347

5. Joseph, L., Wolfson, D., Berger, D. B. (1995). Sample size calculations for binomial proportions via highest posterior density intervals. Statistician **44**, 143–154

6. Katsis, A., Toman, B. (1999). Bayesian sample size calculations for binomial experiments. J. Stat. Plan. Infer. **81**, 349–362

7. Katsis, A. (2001). Calculating the Optimal Sample Size for Binomial populations. Commun. Stat. A **30**, 665–678

8. Pham-Gia, T., Turkkan, N. (1992). Sample size determination in Bayesian analysis. Statistician **41**, 389–397

On the Second Order Minimax Improvement of the Sample Mean in the Estimation of a Mean Value of the Exponential Dispersion Family

Zinoviy Landsman

Department of Statistics, University of Haifa, Israel

Abstract. We consider the problem of the second order minimax improvement of the sample mean in the estimation of the mean value of the Exponential Dispersion Family (EDF), when the space of all possible values of mean is nonrestricted. We show a necessary and sufficient conditions for the possibility of such an improvement.

1   Introduction

This paper represents a brief review of the results given in [12].

We study the problem of second order minimax estimation of the mean value, μ, of the Exponential Dispersion Family (EDF)

$$dP_{\theta,\lambda} = e^{\lambda(x\theta - k(\theta))}dQ_\lambda(x), \ \theta \in \Theta \subset R^1, \lambda \in R^+,$$

where θ is a canonical parameter, λ is a dispersion parameter. For references on the EDF see [6], [7], [8], [9],[21] and [22]. EDF includes many standard families such as Normal, Gamma, Inverse Gaussian, Stable and others. We suppose that the Normal Exponential Family (NEF) which generates the EDF (see [9], Sect. 3.1) is regular (see [3], Ch.3) or at least steep. Then the mean function by Theorem 3.6 of [3]

$$\mu = \mu(\theta) = E_{\theta,\lambda} X = \int x dP_{\theta,\lambda} = k'(\theta)$$

is one-to-one (X is a random variable distributed according to $P_{\theta,\lambda}$). The variance is

$$V_{\theta,\lambda} \ X = \frac{1}{\lambda}k''(\theta) = V(\mu)\sigma^2, \tag{1}$$

where $V(\mu)$ is called the *variance function*.

Let $X_1, ..., X_n$ be i.i.d. observations from the EDF. Then $\bar{X}_n = 1/n \sum_{i=1}^n X_i$, being the maximum likelihood estimator (MLE) of μ, is a first order optimal estimator. We focus attention on the problem of improving the efficiency of $\bar{X}_n$ in the second order for a nonrestricted space of all possible values of μ. We show that this improvement depends on the property of a pair of measures, determined by the model variance function $V(\mu)$ and some weight function $q(\mu)$, to be a *strong type pair* for an integral operator. This property, investigated in [1], [2], [19] (Sect. 1.3.1), complements and generalizes a well-known Hardy inequality (see [4]). Applying their results we obtain a necessary and sufficient condition for such a possibility, and as an application we test the Tweedie model, one of the important submodels of the EDF, which contains the most popular distributions: Normal, Gamma, Inverse Gauss and others.

In Section 2 we study the asymptotic properties of generalized second order Bayes estimators of the mean in the case of the EDF. In Section 3 we analyze the second order minimaxity property on unrestricted intervals. Section 4 is devoted to the application of the previous results to Tweedie EDF models.

Current interest in estimating the mean value, μ, is due to the attractive role the EDF plays in actuarial science, in the context of credible estimation of μ,"credibility formula" (see [5], [10] (Sect. 5), [15], [16]).

2   Second Order Generalized Bayes Estimator of the Mean

Generalized Bayes estimator (g.B.e.) is the main tool in the minimax investigation. g.B.e. of the mean parameter $\mu = k'(\theta)$, in the case of NEF, was considered in ([3], Ch. 4). In this section we obtain the asymptotics of g.B.e. and their risks up to the term $O(\frac{1}{n})$, defined for the quadratic loss, for EDF. Our treatment, taking into account the definition of EDF and based on the asymptotics of a well-known Laplace integral, gives the asymptotics of generalized Bayes estimator under weaker conditions than those in [13] and [18]. For random variable Y_n we write $Y_n = o_{L_2}(\alpha_n)$ if $EY_n^2 = o(\alpha_n)$.

Theorem 1. *Let $\pi(\theta)$ be a generalized prior density and $\pi(\theta)\exp(-k(\theta))$ be absolutely continuous on R. Suppose that for $\lambda \in R_+$ and $x \in S$*

$$\int_\Theta \exp(\lambda(\theta x - k(\theta)))\pi(\theta)d\theta < \infty, \tag{2}$$

$$\int_\Theta |\pi'(\theta)| \exp(\lambda(\theta x - k(\theta)))d\theta < \infty, \tag{3}$$

$$\int_\Theta |k'(\theta)| \exp(\lambda(\theta x - k(\theta))\pi(\theta)d\theta < \infty. \tag{4}$$

Let $supp(\pi(\theta)) = [a,b] \subset \Theta$ for some $-\infty \leq a < b \leq \infty$. Then the generalized Bayes estimator of $\mu = \mu(\theta), \theta \in (a,b)$ has the asymptotic expansion

$$\mu_n = \bar{X}_n + \frac{1}{n\lambda}V(\mu)\frac{\tilde{\pi}'(\mu)}{\tilde{\pi}(\mu)}|_{\mu=\bar{X}_n} + o_{L_2}(\frac{1}{n}), \tag{5}$$

where $\tilde{\pi}(\mu(\theta)) = \pi(\theta)$, and this asymptotic expansion is uniform with respect to θ on any finite subinterval of $[a, b]$.

The proves of this and following theorems are given in Landsman ([12]).

Definition 1. . The estimator

$$\tilde{\mu}_n = \bar{X}_n + \frac{1}{n\lambda}V(\mu)\frac{\tilde{\pi}'(\mu)}{\tilde{\pi}(\mu)}|_{\mu=\bar{X}_n},$$

which coincides with the generalized Bayes estimator μ_n up to the term $o_{L_2}(\frac{1}{n})$ (see (5)), is called the second order generalized Bayes estimator.

Defining

$$\omega(\mu) = \sqrt{\tilde{\pi}(\mu)} \tag{6}$$

and substituting (6) into the first formula in Definition 1 we can redesignate $\tilde{\mu}_n$ by $\mu_{n,\omega}$ and give its representation in terms of ω as

$$\mu_{n,\omega} = \bar{X}_n + \frac{2}{n\lambda}V(\mu)\frac{\omega'(\mu)}{\omega(\mu)}|_{\mu=\bar{X}_n} .$$

The risk function of $\mu_{n,\omega}$ can be given in terms of ω by the following:

Theorem 2. . In addition to the assumptions of Theorem 1, let $\tilde{\pi}(\mu)$ be three times continuously differentiable on $[\tilde{a}, \tilde{b}]$. Then for $\mu \in (\tilde{a}, \tilde{b})$

$$E_{\mu,\lambda}(\mu_{n,\omega} - \mu)^2 = \frac{1}{\lambda n}V(\mu) + \frac{V(\mu)}{n^2\lambda^2}\frac{L\omega}{\omega(\mu)} + o(\frac{1}{n^2})$$

uniformly in μ on any closed subinterval of $(\tilde{a}, \tilde{b})$, where

$$L\omega = 4\frac{d}{d\mu}V(\mu)\frac{d}{d\mu}\omega(\mu) \tag{7}$$

is the Sturme-Liouville differential operator.

3   Second Order Minimaxity on the Infinite Interval of Values of μ.

Let the set of all possible values of the mean $\mu, K = k'(\Theta)$, be either the positive half line of R or R. Let $q(\mu) > 0$ be some weight function. Define the second order minimax constant with respect to q

$$\delta(q, K) = \lim_{n\to\infty} \inf_{\mu_n} \sup_{\mu\in K} n^2 q(\mu)^{-1}(E_{\mu,\lambda}(\mu_n - \mu)^2 - \frac{1}{\lambda n}V(\mu)). \tag{8}$$

Here μ_n is any estimator of μ, and $E_{\mu,\lambda}(\cdot) = E_{\theta(\mu),\lambda}(\cdot)$. Let us say that μ_n^* is a second order minimax estimator of μ with respect to the weight $q(\mu)$, if for any $\mu \in K$

$$\lim_{n\to\infty} n^2 q(\mu)^{-1}(E_{\mu,\lambda}(\mu_n^* - \mu)^2 - \frac{1}{\lambda n}V(\mu)) \le \delta(q,K).$$

Here and further on we assume that the weight function $q(\cdot)$ is in $C^1(K)$. Special cases of weight functions are $q(\mu) = 1$ or $q(\mu) = V(\mu)/\lambda$. The last one reduces $\delta(q,K)$ to the following form

$$\delta(q,K) = \lim_{n\to\infty} \inf_{\mu_n} \sup_K n(\frac{E_{\mu,\lambda}(\mu_n - \mu)^2}{V_{\mu,\lambda}(\bar{X}_n)} - 1).$$

It is clear that

$$\delta(q,K) \le 0$$

and μ_n^* yields a smaller risk than $\bar{X}_n$, if

$$\delta(q,K) < 0. \qquad (9)$$

Let us first consider the case $K = R_+$. We say, following [1], that a pair of measures (ν,γ) defined on set K is a *strong type pair* for the linear operator $T : L_2(\gamma,K) \to L_2(\nu,K)$, if there exists a constant C, independent on f, such that

$$(\int_K |Tf(x)|^2 d\nu)^{1/2} \le C(\int_K |f(x)|^2 d\gamma)^{1/2}. \qquad (10)$$

Here $L_2(\gamma,K)$ and $L_2(\nu,K)$ are Hilbert spaces of functions which are square integrable on K with respect to measures γ and ν respectively. The smallest choice of the constant C is called the strong norm of the operator T and is denoted by $\|T\|_s$. We show that (9) is related to the property that the pair of measures on K, (ν,γ), defined by the model variance function $V(\mu)$ and a weight function $q(\mu)$ as

$$d\nu = q(x)V(x)^{-1}dx, d\gamma = V(x)dx \qquad (11)$$

is a strong type pair for the integral operator $P_0 f(x) = \int_0^x f(t)dt$ or for its dual operator $Q_0 f(x) = \int_x^\infty f(t)dt$.

Suppose $\Delta_l, l = 1, 2\ldots$ is a sequence of bounded intervals in K such that

$$\Delta_l \uparrow K, \text{ as } l \to \infty.$$

With this sequence let us associate the following sequence of positive numbers $\alpha_1(\Delta_l)$, being the first eigenvalues of the corresponding sequence of Dirichlet problems

$$\begin{cases} L\omega + \alpha \frac{q(x)}{V(x)}\omega(x) = 0 \\ \omega(x)_{|\partial\Delta_l} = 0 \end{cases} \qquad (12)$$

with L given in (7).

For a regular or steep EDF, the variance function $V(x)$ is positive on K. Then the Dirichlet problem (12) has the first eigenvalue $\lambda_1(\Delta_l) > 0$. The well-known Dirichlet principle (see e.g. [11], Ch.3, Sec. 17) says that

$$\alpha_1(\Delta_l) = 4 \inf_{u(x) \in H_0^1(\Delta_l)} \frac{\int_{\Delta_l} V(x)u'(x)^2 dx}{\int_{\Delta_l} q(x)V(x)^{-1}u(x)^2 dx} \tag{13}$$

and the infimum is attained by the corresponding to $\alpha_1(\Delta_l)$ eigenfunction $\omega_{\Delta_l}(x)$, which is positive on Δ_l and smooth because $V(x)$ and $q(x)$ are smooth. As $\Delta_l, l = 1, 2, ..$ is a monotone sequence of intervals, $\alpha_1(\Delta_l), l = 1, 2, ...$ is a non-increasing sequence of positive numbers. Let

$$\alpha_1 = \lim_{l \to \infty} \alpha_1(\Delta_l). \tag{14}$$

Theorem 3. *If the pair of measures* (ν, γ), *defined in (11), is a strong type pair for the integral operator* P_0, *i.e.* $\|P_0\|_s < \infty$, *then*

$$\delta(q, K) \geq -\frac{\alpha_1}{\lambda^2} \tag{15}$$

and $\alpha_1 = 4/\|P_0\|_s^2 > 0$. *If* $\|P_0\|_s^2 = \infty$, *but* (ν, γ) *is a strong type pair for a dual operator* Q_0, *then* $\alpha_1 = 4/\|Q_0\|_s^2 > 0$, *otherwise* $\alpha_1 = 0$.

Theorem 4. *Let the two measures* ν, γ *be defined by (11), and let*

$$B = \min\{\sup_{r>0}(\nu([r, \infty)) \int_0^r V(x)^{-1} dx),$$
$$\sup_{r>0}(\nu((0, r]) \int_r^\infty V(x)^{-1} dx)\}.$$

Then, if $B = \infty$, $\bar{X}_n$ *is a second order minimax estimator of mean* μ. *If* $B < \infty$, α_1 *in (14) is positive, and the estimator*

$$\mu_{n,\omega^*} = \bar{X}_n + \frac{2}{n\lambda} V(\mu) \frac{\omega^*(\mu)'}{\omega^*(\mu)}|_{\mu=\bar{X}_n},$$

where $\omega^*(x)$ *is a positive solution of equation*

$$L\omega + \alpha_1 \frac{q(x)}{V(x)}\omega(x) = 0, x \in K, \tag{16}$$

is a second order minimax estimator of μ *and*

$$\lim_{n \to \infty} n^2 q(\mu)^{-1}(E_{\theta, \lambda}(\mu_n^* - \mu)^2 - \frac{1}{\lambda n}V(\mu)) = -\frac{\alpha_1}{\lambda^2}.$$

4   Tweedie Models

In this section we apply the previous results to the important subclass of EDF - Tweedie models, for which

$$V(\mu) = \mu^p, \mu \in K_p, \quad p \in P \subset R$$

(see [9], Ch. 4). Many popular distributions, such as Normal, Gamma, Inverse Gauss, are members of the Tweedie family with $(p = 0, K_p = R), (p = 2, K_p = R_+), (p = 3, K_p = R_+)$ respectively. A full description of Tweedie-EDF is given in [9], Table 4.1.

As the variance function V is a power function of μ, it is natural to investigate a second order minimax problem with respect to the power weight function

$$q(\mu) = \mu^t > 0, \ \mu \in K_p. \tag{17}$$

Theorem 5. *Let the classification parameter p for a Tweedie EDF satisfy $p \neq 1$, and $K_p = R_+$. Then $\bar{X}_n$ can be improved in the minimax sense in second order with respect to the power weight $q(\mu) = \mu^t$ iff $t = 2(p-1)$; the second order minimax estimator, which improves $\bar{X}_n$ for any $\mu \in R_+$, is*

$$\mu_n^* = \bar{X}_n(1 - \frac{p-1}{n\lambda}\bar{X}_n^{p-2}) \tag{18}$$

and the risk of μ_n^* is

$$E_{\mu,\lambda}(\mu_n^* - \mu)^2 = \frac{1}{\lambda n}\mu^p(1 - \frac{(p-1)^2}{n\lambda}\mu^{p-2} + o(\frac{1}{n})). \tag{19}$$

$K_p = R$ *is the only case of a normal distribution, then $p = 0$, and $\bar{X}_n$ cannot be improved in second order on whole R for any q.*

Example 1. Gamma distribution with shape parameter α and scale parameter β is EDF with $\mu = \alpha\beta$ and $\lambda = \alpha$. It is a member of the Tweedie class with $p = 2, K_p = R_+$. Theorem 5 says that

$$\mu_n^* = \bar{X}_n(1 - \frac{1}{n\lambda})$$

is a second order minimax estimator, which uniformly reduces the second order of the relative risk, i.e.,

$$\frac{E_{\mu,\lambda}(\mu_n^* - \mu)^2}{V_{\mu,\lambda}(\bar{X}_n)} = 1 - \frac{1}{n\lambda} + o(\frac{1}{n}), n \to \infty.$$

Example 2. The Inverse Gauss distribution belongs to the Tweedie model with $p = 3, K_p = R_+$. From Theorem 5 we have that

$$\mu_n^* = \bar{X}_n(1 - \frac{2}{n\lambda}\bar{X}_n)$$

is a second order minimax estimator with the relative risk

$$\frac{E_{\mu,\lambda}(\mu_n^* - \mu)^2}{V_{\mu,\lambda}(\bar{X}_n)} = (1 - \frac{4}{n\lambda}\mu + o(\frac{1}{n})).$$

References

1. Andersen, K.F., Muckenhoupt, B. (1982). Weighted weak type Hardy inequalities with application to Hilbert transforms and maximal functions. Stud. Math. **72**, 9–26
2. Bradley, J.S. (1978). Hardy inequalities with mixed norms. Can. Math. Bulletin **21**, 405–408
3. Brown, L.D. (1986). *Fundamentals of Statistical Exponential Families*. Lecture Notes - Monograph series, Hayward, CA
4. Hardy, G., Littlewood, J., Polya, G. (1934). *Inequalities*. Cambridge Univ. Press, Cambridge
5. Jewell, W.S. (1974). Credible means are exact Bayesian for exponential families. ASTIN Bull. **8**, 77–90
6. Jorgensen, B. (1986). Some properties of exponential dispersion models. Scand. J. Stat. **13**, 187–198
7. Jorgensen, B. (1987). Exponential dispersion models (with discussion). J. Roy. Stat. Soc. B. **49**, 127–162
8. Jorgensen, B. (1992). Exponential dispersion models and extensions: A review. Int. Stat. Rev. **60**, 5–20
9. Jorgensen, B. (1997). *The Theory of Dispersion Models*. Chapman and Hall, London
10. Klugman, S.A., Panjer, H.H., Willmot, G.E. (1998). *Loss Models. From Data to Decisions*. Wiley, New York
11. Ladyzenskaya, O.A., Uralt'seva, N.N. (1968). *Linear and Quasilinear Elliptic Equations*. Academic Press, New York
12. Landsman, Z. (2001). Second order minimax estimation of the mean value for Exponential Dispersion models. J. Stat. Plan. Infer. **98**, 57–71
13. Landsman, Z., Levit, B. (1990). The second order minimax estimation: nuisance parameters and hypoelliptic operators. Probability Theory and Mathematical Statistics. *Proceedings of the Fifth International Vilnius Conference*. **2**, 47–58, VSP, Utrecht
14. Landsman, Z., Levit, B. (1991). Second order asymptotic minimax estimation in the presence of a nuisance parameter. Probl. Inf. Transm. **26**, 226–244
15. Landsman, Z., Makov, U. (1998). Exponential dispersion models and credibility. Scand. Actuarial J. **1**, 89–96
16. Landsman, Z., Makov, U. (1999). Credibility evaluations for exponential dispersion families. Insur. Math. Econ. **24**, 33–39
17. Levit, B. (1980). Second order minimaxity. Theor. Probab. Appl. **25**, 561–576
18. Levit, B. (1982). Second order minimax estimation and positive solutions of elliptic equations. Theor. Probab. Appl. **27**, 525–546
19. Maz'ja, V.G. (1985). *Sobolev Spaces*. Springer-Verlag, Berlin
20. Nelder, J.A., Verrall, R.J. (1997). Credibility theory and generalized linear models. ASTIN Bull. **27**, 71–82
21. Nelder, J.A., Wedderburn, R.W.M. (1972). Generalized Linear Models. J. Roy. Stat. Soc. A **135**, 370–384
22. Tweedie, M.C.K. (1984). An index which distinguishes between some important exponential families. In: *Statistics: Applications and new directions. Proceedings of the Indian Statistical Golden Jubilee International Conference*. Ghosh, J.K., Roy, J. (Eds.) 579–604. Indian Statistical Institute.

Bayesian Analysis of Cell Migration – Linking Experimental Data and Theoretical Models

Roland Preuss[1], Albrecht Schwab[2], Hans J. Schnittler[3] and Peter Dieterich[3]

[1] Centre for Interdisciplinary Plasma Science, Max-Planck-Institute for Plasmaphysics, Garching, Germany
[2] Institute for Physiology, University of Würzburg, Germany
[3] Institute for Physiology, Technical University of Dresden, Germany

Abstract. We analyze experimental time series from phase contrast microscopy of cells moving on a 2D substrate. Using Bayesian analysis a statistical model is developed which allows to characterize cell migration with a few parameters.

1 Introduction

Cell migration plays a key role in many medical questions, as for example during wound healing and the transmigration of leukocytes or tumor cells. However, it shows to be a highly complex process involving the cooperative interaction of a large variety of biomolecular components. Up to now mathematical models were mainly developed on two scales of description. On the molecular scale, migration is correlated with many biochemical reactions in the cell, i.e. the polymerization and depolymerization of actin or iontransport across cell membranes. In general, this approach could provide full description, though at the end it may be far too complicated and one is in danger of losing the complete picture. Contrarily, on a larger scale, it is focused on the movement of the center of mass of the cell only, e.g. with simple stochastic models. Unfortunately this phenomenological approach may prove to be too crude to allow for differentiated statements about the cell behavior.

We therefore want to pursue a 'middle way' in between the above approaches. We analyze experimental time series from phase contrast microscopy of cells moving on a 2D substrate. These data do not allow to give insight on the microscopic level, but deliver far more information than only the center of mass, e.g. they take account of subcellular processes like the dynamics of protrusions and adhesion releases. Considering this information content a model is developed which allows to characterize cell migration with a few parameters.

The analysis is performed using Bayesian probability theory. Within this framework it is possible to determine which of the above models describe the data best. Though more sophisticated models with larger parameter sets easily fit the data better, the experimental quality may not allow to draw such complicate conclusions and already a simpler model may take full account of the information in the data. Bayesian model comparison automatically includes this principle, called Occam's Razor. The model parameters themselves are evaluated in the form of expectation values over the posterior distribution. Both the model comparison and the parameter estimation are performed numerically involving computation methods like Markov chain Monte Carlo.

2 The Problem

Nearly all biological cells are 'on the move' during their live cycle. In addition, cell migration is of central medical relevance. Examples are embryo genesis, the spreading of tumor cells, transmigration of leukocytes, and wound healing. Thus, it is important to look for criteria to find and quantify changes of cell migration.

However, as schematically shown in Fig. 1 cell migration is a highly correlated process, where a huge and even unknown number of molecular components contributes to a physically coordinated motion of the whole cell [1].

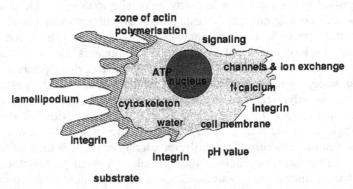

Fig. 1. The cell is a complex system with many interacting components. Some of them which are known to be involved in the machinery of cell migration are labeled with names.

How to gain insight in such complex systems? The system is far too complicated to calculate the dynamics of molecular processes from first principles, e.g. from microscopic laws as the Schrödinger equation. In addition, hierarchical (temporal and spatial) organization on various scales might disallow this in principle [2]. Furthermore modification of components takes place at molecular (microscopic) level, whereas the medical relevance occurs at macroscopic lengths, e.g. a cell missing a certain type of protein can show a completely different migration behavior.

Thus experimental information is necessary to find and model the essential components during cell migration. A typical experiment is shown in Fig. 3. Cells are moving on a 2D-substrate and can be observed with microscopic techniques. Fig. 3 shows a typical result, where snapshots of a single cell are displayed at different times together with the segmented cell boundary.

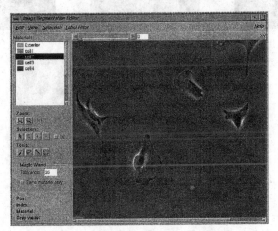

Fig. 2. The cell images are analyzed with a border detection method.

A closer look at the time-lapse series leads to several (known) observations:

- Cells change their state between resting and migrating in a nearly time-periodic way.
- Moving cells are elongated, resting (unpolarized) cells show circular structures.
- Extrusions, so-called lamellipodia, try to pull cells forward.

Many additional phenomena have been observed and can provide an experimental based starting point of modeling the system. Bayesian approach offers criteria to select the model which is most appropriate with respect to the experimental results.

3 The Model

Time-lapse observations of migrating cells deliver information about the cell contours as shown in Fig. 2. From these contours further parameters can be derived, e.g. cell centers, area changes, positions of extrusions. From the observation of time-lapse series a simple model for cell migration is constructed in the following way.

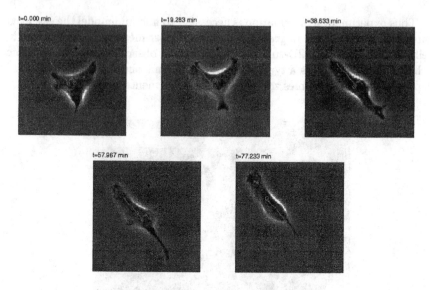

Fig. 3. Migrating cell on a 2D-substrate.

Extrusions of the cell membrane, so-called lamellipodia, generate forces F which drive a cell of mass m and lead to a complex motion of the cell center. The cell is adhesive to the substrate with adherence coefficient β, which captures the complex and dynamical cell-substrate interaction (especially via integrins) in a mean single parameter.

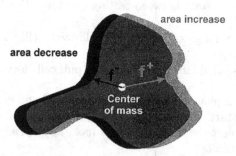

Fig. 4. Cell displacement after one time slice. The light area is new, the dark area vanishes. Summing up all light (dark) pixels at the cell border results in the force f^+ (f^-).

The forces are derived from the time series of cell area displacement (see Fig. 4) in two different ways:

Model 1: The force results from the sum of all vectors going from the center of mass to point of area increases: $F_1 = f^+$

Model 2: In addition to F_1 the vector sum from the center of mass to points of area decreases is taken into account: $F_2 = f^+ - f^-$

Whereas model 1 only captures the pulling forces of lamellipodia, model 2 in addition takes into account the effect of retraction releases of the trailing cell part. The resulting equation of motion is

$$m\frac{\partial^2}{\partial t^2}r + \beta\frac{\partial}{\partial t}r = \gamma F \tag{1}$$

with $r_i = (x_i, y_i)^T$. This constitutes a second-order differential equation which can be transformed by substitution of

$$v = \frac{\partial}{\partial t}r \tag{2}$$

into

$$\frac{\partial}{\partial t}v = -bv + cF \tag{3}$$

where $b = \beta/m$ and $c = \gamma/m$. Eq. (2) and Eq. (3) are two first-order differential equations and can be easily accessed with numerical methods like Runge-Kutta.

4 Bayesian Formalism

4.1 Parameter Estimation

The data consist of the (x,y)-positions of the cell center for every time slice i, which is determined from experiment with measurement error ε.

$$d_i = r_i(b, c) + \varepsilon \tag{4}$$

With the assumption of $\langle \varepsilon \rangle = 0$ and $\langle \varepsilon^2 \rangle = \sigma^2$ we get by the maximum entropy principle [3] a Gaussian likelihood

$$p(D|b, c, I) = \frac{1}{(2\pi\sigma^2)^{\frac{N}{2}}} \exp\left\{-\frac{1}{2\sigma^2} \sum_{i=1}^{N} |d_i - r_i(b, c)|^2\right\} \tag{5}$$

The following prior for the parameters is obtained again by invoking the maximum entropy principle, where we use a first estimate b_0 and c_0 derived from other experiments and principle considerations as a constraint

$$p(b|b_0, I) = \frac{1}{b_0} \exp\left\{-\frac{b}{b_0}\right\} \quad , \quad p(c|c_0, I) = \frac{1}{c_0} \exp\left\{-\frac{c}{c_0}\right\} \tag{6}$$

The two parameters b and c are determined in the form of expectation values

$$\langle b \rangle = \frac{\int b\, p(b|D, I)\, db}{\int p(b|D, I)\, db} = \frac{\int b\, p(b, c|D, I)\, db\, dc}{\int p(b, c|D, I)\, db\, dc} \tag{7}$$

With the help of Bayes theorem

$$p(b, c|D, I) = \frac{p(b, c|I)}{p(D|I)} p(D|b, c) \tag{8}$$

and inserting (5) and (6) we get

$$\langle b \rangle = \int b \frac{\exp\left\{-\frac{1}{2\sigma^2}\sum_{i=1}^{N}|d_i - r_i(b, c)|^2 - \frac{b}{b_0} - \frac{c}{c_0}\right\}}{\int \exp\left\{-\frac{1}{2\sigma^2}\sum_{i=1}^{N}|d_i - r_i(b', c')|^2 - \frac{b'}{b_0} - \frac{c'}{c_0}\right\} db' \, dc'} db \, dc \quad . \tag{9}$$

The fraction in Eq. (9) can be regarded as a sampling density for the Markov chain Monte Carlo method in order to calculate the expectation values numerically.

4.2   Model Comparison

In order to compare different models, we are looking for the probability of a model M_k given the data D. Employing Bayes theorem again delivers

$$p(M_k|D, I) = \frac{p(M_k|I)}{p(D|I)} p(D|M_k, I) \quad . \tag{10}$$

Since no model is preferred a priori, $p(M_k|I) = const$. The evidence cancels out in comparing two model probabilities obtained for the same data and we are left with the determination of the global likelihood $p(D|M_k, I)$. The latter is assigned to the probability functions of Eq. (5) and (6) by marginalization of the parameters $\theta^T = (b, c)$:

$$p(D|M_k, I) = \int p(D, \theta|M_k, I) \, d\theta$$

$$= \int p(D|\theta, M_k, I) p(\theta|I) d\theta \quad . \tag{11}$$

The integrand in equation (11) is mainly of Gaussian shape and can be approximated by expanding the exponent $\Phi(\theta)$ around its maximum $\Phi(\theta_0)$ using Laplace approximation ([3]):

$$\Phi(\theta) = \Phi_0 + \frac{1}{2}\Delta\theta^T H \Delta\theta \quad . \tag{12}$$

Now it is possible to perform the integration analytically which results in

$$p(D|M_k, I) = const \cdot \frac{\exp\left\{\Phi(\theta_0)\right\}}{\sqrt{\det H}} \quad . \tag{13}$$

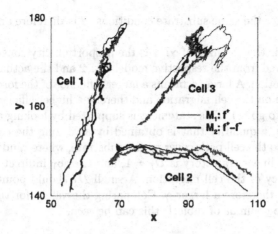

Fig. 5. Resulting cell path for model 1 (thick black line) and model 2 (thin black line compared to the actual path (gray line).

5   Results

For the two models the resulting cell path for the three cells is shown in Fig. 5. Already eye sight tells us that model 2 wins clearly over model 1. The actual calculation shows that the model probability for model 1 is negligible compared to model 2. Further results of the model parameters are therefore only shown for model 2 (see table 1). In order to get β and γ the resulting

Table 1. Parameter values for the three cells under consideration

Cell	A (# pixels)	β (a.u.)	γ (a.u.)	D_f
1	5000	995 ± 100	1000 ± 100	1.17
2	10500	1050 ± 105	630 ± 63	1.32
3	3500	700 ± 70	525 ± 53	1.27

expectation values $\langle b/m \rangle$ and $\langle c/m \rangle$ have to be multiplied with the mass of each cell. Assuming, that the mass of the cell is proportional to the area coverage A which can be determined by counting the pixels within the cell boundaries of Fig. 2, the area delivers a simple estimate of the cell mass. A is given in table 1 where the number of pixels is the unit. Multiplying this value with $\langle b/m \rangle$ and $\langle c/m \rangle$ results in the desired quantities β and γ, however in arbitrary units (a.u.) (see third and forth column of Table 1, an error of 10% is assumed for deviations in $A \propto m$).

Let us first discuss β which reveals more or less the same value for the three cells. This should be the case since all three cells are from the same cell culture and face the same substrate conditions. β is therefore an appropriate parameter.

But what is the meaning of γ? γ is the proportionality constant between the force derived from the respective model 1 or 2 and the actual force in the equation of motion. A larger γ means a larger influence of the force generating lamellipodium on the cell migration and thereby a higher efficiency (the cells knows where to go to). This statement is supported by looking at the fractal dimension D_f, a quantity that is obtained by analyzing the cell path. It is proportional to the cell path covering the substrate, where random motion is $D_f = 2$ and a linear trajectory is $D_f = 1$, and thereby indirect proportional to the efficiency of the cell migration. A small D_f should point to a higher efficiency and therefore a larger γ. Comparing the values for the three cells in the last two columns of table 1, this can be seen.

6   Conclusion

The present work derives simple models describing cell migration with two parameters where driving forces were constructed from areas of increases and decreases. Bayesian model comparison was able to prefer by far the model which includes the effects of driving lamellipodia forces and retractions of trailing ends.

In future we will extend our approach to more distinguished models and apply the technique to a larger number and variety of cells.

Acknowledgements

This work was supported by Deutsche Forschungsgemeinschaft and the Med-Drive program of the Medical Faculty of the University at Dresden. The first author is indebted to V. Dose for discussions about Bayesian analysis solving differential equations.

References

1. Lauffenburger D.A., Horwitz A.F. (1996). Cell migration: a physically integrated molecular process. Cell **84**, 359–369
2. Laughlin, R.B., Pines, D. (2000). The theory of everything. P. Natl. Acad. Sci. USA **97**, 28–31
3. Sivia, D.S. (1996). *Data analysis: a bayesian tutorial.* Oxford University Press, Oxford

Part III

Testing, Goodness of Fit and Randomness

Sequential Bayes Detection of Trend Changes

Martin Beibel[1] and Hans R. Lerche[2]

[1] Research Group, Novartis Pharma AG, Basel, Switzerland
[2] Department of Mathematical Stochastics, University of Freiburg, Germany

Abstract. Let W_t ($0 \leq t < \infty$) denote a Brownian motion process which has zero drift during the time interval $[0, \nu)$ and drift θ during the time interval $[\nu, \infty)$, where θ and ν are unknown. The process W is observed sequentially. The general goal is to find a stopping time T of W that 'detects' the unknown time point ν as soon and as reliably as possible on the basis of this information. We work in a Bayesian framework and discuss a loss structure that is closely connected to that of the Bayes tests of power one of Lerche ([4]). This work extends Beibel's ([2]) where only normal priors on θ were studied. An important ingredient in our proof is the comparison of the process of the posterior variance under different priors similar to the arguments in Paulsen ([6]).

1 Introduction

In the study of financial markets the quick detection of changes of market trends is an important issue for prognosis. The question rises which are good indicators. Of course the answer depends on the setting under study. For instance it has been shown that the Cusum-statistic has a minimax property when one assumes that nature choses the change-point as unfavorable as possible (see Ritov ([7]) and Beibel ([1])). Here we study this problem from a Bayesian point of view in a classical setting. We assume that the market under observation moves at first like a Brownian motion with a known drift. At an unknown random time the drift changes to an unknown size. The task is to detect this change as quickly and as reliable as possible. We solve this problem for rather arbitrary priors of drift size and time of change. In the paper of Beibel ([2]) the distribution of the drift was assumed to be normal. Here we combine the approach of Beibel with ideas of Paulsen ([6]). We consider the following setting. Let $W_t, t \geq 0$ denote a Brownian motion with drift zero up to time ν and drift θ during the time interval $[\nu, \infty)$. We denote by $P_{(\theta,\nu)}$ the corresponding probability measure and by P_∞ the probability measure when no change of the drift occurs. This means that $P_\infty = P_{(\theta,\infty)}$ for all $\theta \in (-\infty, \infty)$ and P_∞ is the measure of standard Brownian motion. Let $E_{(\theta,\nu)}$ denote the expectation with respect to $P_{(\theta,\nu)}$ and E_∞ with respect to P_∞. We look for a stopping time T of W which will stop soon after ν with low probability of alarm under P_∞. Following Lerche ([4]) we study the Bayes risk

$$L(c,T) = P_\infty(T < \infty) + c \int_{-\infty}^{+\infty} \left\{ \frac{\theta^2}{2} \int_{[0,\infty)} E_{(\theta,\nu)}(T-\nu)^+ \rho(d\nu) \right\} G(d\theta),$$

where ρ and G are probability distributions on $[0, \infty)$ and $(-\infty, +\infty)$ respectively. The goal is to minimize $L(c, T)$ over all stopping rules. We provide an asymptotic expansion of the minimal Bayes risk when the costs c tend to zero. We also show that certain mixture stopping rules are asymptotically optimal. For related results when G is normal and for some history see [2].

2   Assumptions and Results

Throughout this article we assume that $\int_0^\infty \nu \rho(d\nu) < \infty$. We further assume that G satisfies

(A1)   The distribution G has an absolutely continuous Lebesgue density g on $(-\infty, \infty)$. That is $G((-\infty, x]) = \int_{-\infty}^x g(y) dy$, where $g(y) = g(0) + \int_0^y g'(z) dz$.

(A2) $\int_{-\infty}^{+\infty} |y|^{2+\delta} g(y) dy < \infty$ for some $\delta > 0$.

(A3) $\int_{-\infty}^{+\infty} \left| H(y) \log |H(y)| \right| g(y) dy < \infty$, where $H(y) = \frac{g'(y)}{g(y)}$.

Let $L_c^* = \inf_T L(c, T)$, where the infimum is taken over all stopping times T of W. Let

$$S_b = \inf \left\{ t > 0 \;\middle|\; \int_{-\infty}^{+\infty} \int_0^\infty e^{y(W_t - W_{t \wedge s}) - \frac{y^2}{2}(t-s)^+} \rho(ds) g(y) dy > b \right\}$$

and $\beta(c) = 1/c$.

Theorem 1 If G satisfies (A1) to (A3), then
$L\left(c, S_{\beta(c)}\right) = c\left(\log \frac{1}{c} + \log\log \frac{1}{c} + K(\rho, G)\right) + o(c)$ as $c \to 0$. The constant $K(\rho, G)$ is given in detail in section 4.

Theorem 2 If G satisfies (A1) to (A3), then $L_c^* = L\left(c, S_{\beta(c)}\right) + o(c)$ as $c \to 0$.

We note that the stopping time S_b has a simple interpretation. Namely, one stops as soon as the mixture of likelihoods

$$\int_0^\infty \int_{-\infty}^{+\infty} e^{y(W_t - W_{t \wedge s}) - \frac{y^2}{2}(t-s)^+} g(y) dy \rho(ds)$$

exceeds b. In the one-sided case, when $\Theta = [0, \infty)$ similar results hold (see [3]). The proofs of the corresponding results are a little more difficult than in the two-sided case. This is due to the possible discontinuity of g at zero. Instead of (A1)-(A3) one can assume that (A1') to (A3') hold, which are stated as following:

(A1')   The distribution G has an absolutely continuous Lebesgue density g on $[0, \infty)$. That is $G((-\infty, 0]) = 0$ and $G((0, x]) = \int_0^x g(y) dy$, where $g(y) = g(0) + \int_0^y g'(z) dz$.

(A2') $\int_0^{+\infty} y^{2+\delta} g(y) dy < \infty$ for some $\delta > 0$.

(A3') $\int_0^{+\infty} \left| H(y) \log |H(y)| \right| g(y) dy < \infty$, with $H(y) = \frac{g'(y)}{g(y)}$.

3 Rewriting the Bayes Risk

Let B denote a standard Brownian motion. Let Y be a random variable with $P(Y \leq y) = \int_{-\infty}^{y} G(ds)$ and τ be a nonnegative random variable with $P(\tau > t) = \int_{(t,\infty)} \rho(ds)$ for all $t \geq 0$. Let B, Y and τ be independent. We assume further that G satisfies condition (A1) to (A3). The process which is observed is

$$W_t = B_t + Y(t - \tau)^+ = B_t + \int_0^t R_s ds ,$$

with $R_s = Y1_{\{\tau \leq s\}}$. The distribution of W is given by $P = \int_{-\infty}^{+\infty} \int_{[0,\infty)} P_{(\theta,\nu)} \rho(d\nu) g(\theta) d\theta$. Let

$$\psi_t = \int_{-\infty}^{+\infty} \int_0^\infty e^{y(W_t - W_{t \wedge s}) - \frac{y^2}{2}(t-s)^+} \rho(ds) g(y) dy.$$

Let $\mathcal{F}_t = \mathcal{F}_t^W = \sigma(W_s; 0 \leq s \leq t)$. Obviously it holds

$$\psi_t = \left. \frac{dP}{dP_\infty} \right|_{\mathcal{F}_t} .$$

Let P_∞ denote the probability measure under which W is a standard Brownian motion. Now we have:

$$L(c,T) = P_\infty(T < \infty) + cE\left\{ \frac{Y^2}{2}(T - \tau)^+ \right\} . \tag{1}$$

To rewrite this risk we derive a stochastic differential equation for $\log \psi_t$ in terms of observable quantities. Let

$$\widehat{R}_t = E(R_t | \mathcal{F}_t) = \psi_t^{-1} \int_{[0,t]} \int_{-\infty}^{+\infty} y e^{y(W_t - W_{t \wedge s}) - \frac{y^2}{2}(t-s)^+} g(y) dy \rho(ds)$$

and

$$\widehat{R_t^2} = E(R_t^2 | \mathcal{F}_t) = \psi_t^{-1} \int_{[0,t]} \int_{-\infty}^{+\infty} y^2 e^{y(W_t - W_{t \wedge s}) - \frac{y^2}{2}(t-s)^+} g(y) dy \rho(ds).$$

Let $\overline{W}$ denote the innovation process

$$\overline{W}_t = W_t - \int_0^t \widehat{R}_s ds .$$

This process is a standard Brownian motion under the probability measure P relative to the filtration $\mathcal{F}$ (see Liptser and Shiryaev ([5]), 297-299). We now obtain:

Proposition 1

$$d\log\psi_t = \frac{1}{2}\left(\hat{R}_t\right)^2 dt + \hat{R}_t d\overline{W}_t \ . \tag{2}$$

Proof. Let

$$Z_t^\infty = \exp\left\{-\int_0^t \hat{R}_s d\overline{W}_s - \frac{1}{2}\int_0^t \left(\hat{R}_s\right)^2 ds\right\} \ .$$

Since $E\int_0^t |R_s| ds < \infty$ for all $t > 0$, we obtain from Wong and Hajek ([9], 254–257) that $\psi_t = Z_t^\infty$.

We will now use Proposition 1 to rewrite $L(c,T)$. Fubinis theorem yields for all $\mathcal{F}^W$-stopping times T

$$E\left(\int_0^T \widehat{R_s^2} ds\right) = E\left\{Y^2(T-\tau)^+\right\} \ .$$

Moreover by a standard likelihood-ratio argument of sequential statistics $P_\infty(T < \infty) = E(1/\psi_T 1_{\{T<\infty\}})$. Therefore we have for all $\mathcal{F}^W$-stopping times T with $L(c,T) < \infty$

$$L(c,T) = E\left\{\frac{1}{\psi_T} + \frac{c}{2}\int_0^T \widehat{R_s^2} ds\right\} \ . \tag{3}$$

Combining Proposition 1 and Equation (3) yields the following representation.

Proposition 2 *For all stopping times T with $E(Y^2(T-\tau)^+) < \infty$ it holds*

$$E\left(\frac{Y^2}{2}(T-\tau)^+\right) = E\left[\log\psi_T + \frac{1}{2}\int_0^T \left(\widehat{R_s^2} - \left(\hat{R}_s\right)^2\right) ds\right]$$

and

$$L(c,T) = E\left[\frac{1}{\psi_T} + c\log\psi_T + \frac{c}{2}\int_0^T \left(\widehat{R_s^2} - \left(\hat{R}_s\right)^2\right) ds\right] \ .$$

In the next sextion we study the third term on the right-hand side.

4   The Integrated Posterior Variance

Let V_t denote the integrated posterior variance (up to a factor $\frac{1}{2}$)

$$V_t = \frac{1}{2}\int_0^t \left(\widehat{R_s^2} - \left(\hat{R}_s\right)^2\right) ds. \tag{4}$$

To study $E(V_t)$ we need some further notation. Let $\mathcal{F}_t^{W,\tau}$ denote the σ-algebra $\sigma(W_s; 0 \leq s \leq t, \tau)$. The filtration $\mathcal{F}^{W,\tau}$ corresponds to a hypothetical observer who knows τ in advance but is still ignorant about the magnitude of the drift Y. Let

$$\hat{R}_t^{(\tau)} = \frac{\int_{-\infty}^{+\infty} y e^{y(W_t - W_\tau) - \frac{y^2}{2}(t-\tau)} g(y) dy}{\int_{-\infty}^{+\infty} e^{y(W_t - W_\tau) - \frac{y^2}{2}(t-\tau)} g(y) dy} 1_{\{\tau \leq t\}} = \mu(W_t - W_\tau, t - \tau) 1_{\{\tau \leq t\}},$$

with

$$\mu(x,t) = \frac{\int_{-\infty}^{+\infty} y e^{yx - \frac{y^2}{2}t} g(y) dy}{\int_{-\infty}^{+\infty} e^{yx - \frac{y^2}{2}t} g(y) dy} .$$

We have $E(R_t | \mathcal{F}_t^{W,\tau}) = \hat{R}_t^{(\tau)}$. Let

$$\overline{W}_t^{(\tau)} = W_t - W_{t \wedge \tau} - \int_0^t \hat{R}_t^{(\tau)}.$$

The process $\overline{W}_t^{(\tau)}$ is a Brownian motion with respect to the filtration $\mathcal{F}_t^{W,\tau}$ for $\tau \leq t < \infty$ starting at $\overline{W}_\tau^{(\tau)} = 0$. Let

$$\widehat{R_t^2}^{(\tau)} = \frac{\int_{-\infty}^{+\infty} y^2 e^{y(W_t - W_\tau) - \frac{y^2}{2}(t-\tau)} g(y) dy}{\int_{-\infty}^{+\infty} e^{y(W_t - W_\tau) - \frac{y^2}{2}(t-\tau)} g(y) dy} 1_{\{\tau \leq t\}} = \rho(W_t - W_\tau, t - \tau) 1_{\{\tau \leq t\}},$$

with

$$\rho(x,t) = \frac{\int_{-\infty}^{+\infty} y^2 e^{yx - \frac{y^2}{2}t} g(y) dy}{\int_{-\infty}^{+\infty} y e^{yx - \frac{y^2}{2}t} g(y) dy} .$$

It holds $E(R_t^2 | \mathcal{F}_t^{W,\tau}) = \widehat{R_t^2}^{(\tau)}$. Let $v(x,t) = \rho(x,t) - \mu(x,t)^2$ denote the posterior variance.

Lemma 1

$$E(V_T) = \frac{1}{2} E \int_0^T v(W_s - W_\tau, s - \tau) 1_{\{\tau \leq s\}} ds + \frac{1}{2} E \int_0^T \left(\hat{R}_s^{(\tau)} - \hat{R}_s \right)^2 ds.$$

Proof. It holds that

$$E\left(\left(R_t - \hat{R}_t^{(\tau)} \right)^2 \Big| \mathcal{F}_t^{W,\tau} \right) = v(W_t - W_\tau, t - \tau) 1_{\{\tau \leq t\}} . \tag{5}$$

Since

$$E\left\{ \left(R_t - \hat{R}_t^{(\tau)} \right) \left(\hat{R}_t^{(\tau)} - \hat{R}_t \right) \Big| \mathcal{F}_t^{W,\tau} \right\} = 0$$

a Fubini type of argument yields

$$E \int_0^T \left(\widehat{R_s^2} - \left(\hat{R}_s \right)^2 \right) ds = E \int_0^T \left(R_s - \hat{R}_s^{(\tau)} \right)^2 ds + E \int_0^T \left(\hat{R}_s^{(\tau)} - \hat{R}_s \right)^2 ds .$$

To calculate $E\left(\int_0^T \left(\hat{R}_s^{(\tau)} - \hat{R}_s\right)^2 ds\right)$, we have to introduce one more probability measure on $\sigma(W_s; 0 \le s < \infty, Y, \tau)$. Let P_0 denote the probability measure under which the process $(W_t - Yt; 0 \le t < \infty)$ is a standard Brownian motion. The distribution of W under P_0 is given by $\int_{-\infty}^{+\infty} P_{(\theta,0)} G(d\theta)$. We will use the likelihood ratio of P with respect to P_0 relative to the filtrations $\mathcal{F}$ and $\mathcal{F}^{W,\tau}$ later on. The probability measures P and P_0 are equivalent on the σ-algebra $\sigma(W_s; 0 \le s < \infty, \tau)$. Let

$$N_\infty = \left.\frac{dP}{dP_0}\right|_{\sigma(W_s;0\le s<\infty)} \quad \text{and} \quad N_\infty^{(\tau)} = \left.\frac{dP}{dP_0}\right|_{\sigma(W_s;0\le s<\infty,\tau)}.$$

The quantities $E(\log N_\infty)$ and $E(\log N_\infty^{(\tau)})$ appear below (see Equation (8)). We have

$$N_\infty = \int_0^\infty e^{-YW_s+\frac{Y^2}{2}s}\rho(ds) \quad \text{and} \quad N_\infty^{(\tau)} = e^{-YW_\tau+\frac{Y^2}{2}\tau}.$$

It is easy to see that

$$E\left(\log N_\infty^{(\tau)}\right) = \left(\int_{-\infty}^{+\infty} \frac{y^2}{2}g(y)dy\right)\left(\int_0^{+\infty} s\rho(ds)\right) < \infty. \tag{6}$$

Since $\mathcal{F}_t = \sigma(W_s; 0 \le s < \infty) \subset \sigma(W_s; 0 \le s < \infty, \tau) = \mathcal{F}_t^{W,\tau}$ we get

$$E(\log N_\infty) \le E(\log N_\infty^{(\tau)}).$$

Moreover

$$E\left(\log N_\infty\right) \tag{7}$$

$$= \int_{-\infty}^\infty \int_{[0,\infty)} \left\{ E_{(\theta,\nu)} \log \left(\int_{[0,\infty)} e^{-\theta W_s+\frac{\theta^2}{2}s}\rho(ds)\right) \right\} \rho(d\nu)G(d\theta).$$

Similar arguments as in [2], p. 473, with $U_s = \frac{W_s}{(s+1)}$ provide

$$E\left(\int_0^t \left(\hat{R}_s^{(\tau)} - \hat{R}_s\right)^2 ds\right) = E\left(\int_0^t \left(\hat{R}_s^{(\tau)} - U_s\right)^2 ds\right) - E\left(\int_0^t \left(\hat{R}_s - U_s\right)^2 ds\right).$$

$$\frac{1}{2}E\left(\int_0^t \left(\hat{R}_s - U_s\right)^2 ds\right) = E(\log N_t) \quad \text{and}$$

$$\frac{1}{2}E\left(\int_0^t \left(\hat{R}_s^{(\tau)} - U_s\right)^2 ds\right) = E\left(\log N_t^{(\tau)}\right).$$

Therefore

$$\frac{1}{2}E\left(\int_0^\infty \left(\hat{R}_s^{(\tau)} - \hat{R}_s\right)^2 ds\right) = E\left(\log N_\infty^{(\tau)}\right) - E\left(\log N_\infty\right) < \infty. \tag{8}$$

The following Proposition 3 will be proved in the Appendix.

Proposition 3

$$E \int_{\tau}^{+\infty} \left| v(W_s - W_\tau, s - \tau) - \frac{1}{s - \tau + 1} \right| ds < +\infty.$$

We note that in the case of a standard normal prior G, the posterior variance $v(x, t)$ is equal to $1/(t + 1)$. Then $E \left(\int_{\tau}^{T} \frac{1}{s - \tau + 1} ds \right) = E \log \left((T - \tau)^+ + 1 \right)$.

Combining Lemma 1, equation 8 and Proposition 3 then yields

Proposition 4 *For all stopping times T with $E(Y^2 (T - \tau)^+) < \infty$*

$$E(V_T) = \frac{1}{2} E \left[\log \left((T - \tau)^+ + 1 \right) \right]$$

$$+ \frac{1}{2} E \left(\int_{\tau \wedge T}^{T} \left[v(W_s - W_\tau), s - \tau) - \frac{1}{s - \tau + 1} \right] ds \right)$$

$$+ \frac{1}{2} E \left(\int_{0}^{T} \left(\hat{R}_s^{(\tau)} - \hat{R}_s \right)^2 ds \right). \tag{9}$$

Moreover

$$E \int_{\tau \wedge T}^{T} \left[v(W_s - W_\tau, s - \tau) - \frac{1}{s - \tau + 1} \right] ds$$

$$\leq E \int_{\tau}^{\infty} \left| v(W_s - W_\tau, s - \tau) - \frac{1}{s - \tau + 1} \right| ds < \infty$$

and

$$E \int_{0}^{T} (\hat{R}_s^{(\tau)} - \hat{R}_s)^2 ds$$

$$\leq E \int_{0}^{\infty} (\hat{R}_s^{(\tau)} - \hat{R}_s)^2 ds = E \left(\log N_\infty^{(\tau)} \right) - E (\log N_\infty) < \infty.$$

5 Mixture Stopping Rules

We now study the stopping times S_b more closely. The following arguments are similar to those of [2]. We recall that $S_b = \inf\{t > 0 | \psi_t \geq b\}$. We have $P(S_b < \infty) = 1$ for all $b > 1$ since the probabilities P_∞ and P are orthogonal on $\sigma(W_s, 0 \leq s < \infty)$. The event $\cap_{k=1}^{\infty} \{S_k < \infty\}$ has probability 1. Therefore since $S_k = k$ holds, it follows

$$P \left(\lim_{b \to \infty} S_b = +\infty \right) = 1.$$

The next lemma describes the asymptotic behavior of S_b.

Lemma 2

$$P\left(\lim_{b\to\infty}\frac{S_b}{\log b}=\frac{2}{Y^2}\right)=1$$

Proof. We have with probability one

$$\log b = \log\psi_t = \int_0^{S_b}[\hat{R}_s]^2 ds + \int_0^{S_b}\hat{R}_s d\overline{W}_s.$$

The process $\int_0^t \hat{R}_s d\overline{W}_s$ is a time-transformed Brownian motion and therefore

$$P\left(\lim_{t\to\infty}\frac{1}{t}\int_0^t \hat{R}_s d\overline{W}_s = 0\right)=1.$$

Moreover

$$P\left(\lim_{t\to\infty}\frac{W_t}{t}=Y\right)=P\left(\lim_{t\to\infty}\frac{B_t-Y(t\wedge\tau)}{t}=0\right)=1.$$

Hence $P(\lim_{t\to\infty}\hat{R}_t=Y)=1$ and thus

$$P\left(\lim_{t\to\infty}\frac{1}{t}\int_0^t[\hat{R}_s]^2 ds = Y^2\right)=1.$$

The preceeding lemma suggests the following result.

Proposition 5 *As* $b\to\infty$

$$E\left(\log\left((S_b-\tau)^+ +1\right)\right) = \log\log b - \int_{-\infty}^{+\infty}\log\left(\frac{y^2}{2}\right)g(y)dy + o(1)\ .$$

Proof. Let

$$\xi_b = \frac{\frac{Y^2}{2}\{(S_b-\tau)^+ +1\}}{\log b},\qquad E_b=\left\{\xi_b\geq\frac{1}{12}\right\}.$$

It is sufficient to show that $E\log\xi_b = o(1)$ as $b\to\infty$. We split $E\log\xi_b$ into three parts.

i) Let $M>1$. Then, by Lemma 3 (below)

$$E\left(1_{\{\xi_b\geq M\}}\log\xi_b\right)\leq\left(\sup_{b\geq 2}E\xi_b\right)\sup_{x\geq M}\left[\frac{\log x}{x}\right]<\infty$$

and therefore

$$\lim_{M\to\infty}\limsup_{b\to\infty}E\left(1_{\{\xi_b\geq M\}}\log\xi_b\right)=0.$$

ii) On the event E_b^C we have $\xi_b<1/12$ and so by Lemma 4 (below)

$$\limsup_{b\to\infty}E\left(1_{E_b^C}\log\xi_b\right)\leq\log(1/12)\limsup_{b\to\infty}P(E_b^C)=0.$$

Moreover

$$E\left(1_{E_b^C} \log \xi_b\right)$$

$$= (\log\log b)P(E_b^C) + E\left(1_{E_b^C} \log\frac{Y^2}{2}\right) + E\left(1_{E_b^C} \log(S_b - \tau)^+ + 1\right)$$

$$\geq (\log\log b)P(E_b^C) + E\left(1_{E_b^C} \log\frac{Y^2}{2}\right).$$

Therefore $\liminf_{b\to\infty} E\left(1_{E_b^C} \log \xi_b\right) \geq 0$.

iii) The random variable $\log \xi_b$ stays bounded on the event $E_b \cap \{\xi_b \leq M\}$. By Lemma 2 $\xi_b \to 1$ and implies $1_{E_b}1_{\{\xi_b \leq M\}} \log \xi_b \to 0$. Hence

$$E\left(1_{E_b}1_{\{\xi_b \leq M\}} \log \xi_b\right) = 0.$$

Lemma 3 *There exists a constant $A \geq 0$ such that for all $b \geq 1$*

$$E\left(\frac{Y^2}{2}\left(S_b - \tau\right)^+\right) \leq 2\log b + A .$$

Proof. Proposition 2 yields for all positive integers n

$$E\left(\frac{Y^2}{2}\left(S_b \wedge n - \tau\right)^+\right) \leq \log b + \frac{1}{2}E\left(\int_0^{S_b \wedge n}\left(\widehat{R_s^2} - \left(\hat{R}_s\right)^2\right)ds\right) .$$

Proposition 4 now gives

$$E\left(\frac{Y^2}{2}\left(S_b \wedge n - \tau\right)^+\right) \leq \log b + \frac{1}{2}E\log\left(\left(S_b \wedge n - \tau\right)^+ + 1\right) + \tilde{A},$$

where

$$\tilde{A} = E\left(\int_\tau^\infty \left|v(W_s - W_\tau, s - \tau) - \frac{1}{s-\tau+1}\right|ds\right) + E\left(\int_0^\infty (\hat{R}_s^{(\tau)} - \hat{R}_s)^2 ds\right)$$
$$< \infty.$$

Since $\log x \leq x$ and $|E\log(Y^2)| < \infty$, we obtain the assertion.

Lemma 4 *As $b \to \infty$*

$$P\left(\frac{Y^2}{2}(S_b - \tau)^+ \leq \tfrac{1}{12}\log b\right) = o\left(\frac{1}{\log b}\right) .$$

The proof is exactly the same as that of Lemma 4 of [2]. Now we can state Theorem 1 with the precise constant.

Theorem 1:
If G satisfies (A1) to (A3), then

$$L(c, S_{\beta(c)}) = c\left[\log\frac{1}{c} + \frac{1}{2}\log\log\frac{1}{c} + K(\rho, G)\right] + o(c)$$

when c → 0. Here K is given by

$$K(\rho, G) = 1 - \frac{1}{2}\int_{-\infty}^{+\infty}\log\left(\frac{\theta^2}{2}\right)g(\theta)d\theta$$
$$+ E\int_{\tau}^{\infty}\left[v(W_s - W_\tau, s - \tau) - \frac{1}{s - \tau + 1}\right]ds$$
$$+ \left(\int_{-\infty}^{+\infty}\frac{\theta^2}{2}g(\theta)d\theta\right)\left(\int_0^{+\infty}s\rho(ds)\right)$$
$$- \int_{-\infty}^{\infty}\int_{[0,\infty)}\left\{E_{(\theta,\nu)}\log\left(\int_{[0,\infty)}e^{-\theta W_s + \frac{\theta^2}{2}s}\rho(ds)\right)\right\}\rho(d\nu)g(\theta)d\theta.$$

Proof. Proposition 2 yields

$$L(c, S_{\beta(c)}) = E\left(\frac{1}{\psi_{S_{\beta(c)}}} + c\log\psi_{S_{\beta(c)}} + cV_{S_{\beta(c)}}\right).$$

For sufficiently small c it holds that

$$E\left(\frac{1}{\psi_{S_{\beta(c)}}} + c\log\psi_{S_{\beta(c)}}\right) = c + c\log c.$$

From Proposition 5 we obtain

$$E\left(V_{S_{\beta(c)}}\right) = \frac{1}{2}\log\log\frac{1}{c} - \frac{1}{2}\int_{-\infty}^{\infty}\log\left(\frac{y^2}{2}\right)g(y)dy$$
$$+ E\int_0^{\infty}\left[v(W_s - W_\tau, s - \tau) - \frac{1}{s - \tau + 1}\right]1_{\{\tau \le s\}}ds$$
$$+ E\left(\log N_\infty^{(\tau)}\right) - E(\log N_\infty) + o(1),$$

as c → 0. The quantities $E\left(\log N_\infty^{(\tau)}\right)$ and $E(\log N_\infty)$ are evaluated in Equations (6) and (7) above. This yields the assertion.

6   Asymptotic Optimality

We now compare the performance of the stopping times $S_{\beta(c)}$ with the performance of c^2-optimal solutions. This leads to a proof of the asymptotic optimality of the stopping times $S_{\beta(c)}$. The following two lemma correspond to Lemma 6 of [1]. The proof can be found there.

Let $\tilde{S}_c$ for $0 < c \le 1$ be an c^2-optimal stopping rule, that is a stopping time with $L(c, \tilde{S}_c) \le L_c^* + c^2$. We may assume $\tilde{S}_c \le S_{\beta(c)}$. Let $\tilde{\beta}(c) = \beta(c)/(1 - \log c)$.

Lemma 5 *Let $(\tilde{S}_c; 0 < c \le 1)$ be stopping times of W with $L(c, \tilde{S}_c) \le L_c^* + c^2$ and $\tilde{S}_c \le S_{\beta(c)}$. Then*

$$\lim_{c \to 0} E \left(\log \frac{(S_{\beta(c)} - \tau)^+ + 1}{(\tilde{S}_c - \tau)^+ + 1} \right) = 0 .$$

Theorem 2:

$$L_c^* = L\left(c, S_{\beta(c)}\right) + o(c) \quad when \; c \to 0 .$$

Proof. The function $g_c(x) = 1/x + c \log x$ assumes its unique minimum over the interval $(0, 1)$ at $x = \beta(c)$. Proposition 2 and Proposition 4 therefore yield together

$$0 \le L(c, S_{\beta(c)}) - L_c^*$$
$$\le L(c, S_{\beta(c)}) - L(c, \tilde{S}_c) + c^2$$
$$\le E\left\{ \frac{1}{\psi_{S_{\beta(c)}}} + c \log(\psi_{S_{\beta(c)}}) - \frac{1}{\psi_{\tilde{S}_c}} - c \log(\psi_{\tilde{S}_c}) \right\}$$
$$\qquad + cE\left\{ V_{S_{\beta(c)}} - V_{\tilde{S}_c} \right\} + c^2$$
$$\le cE\left\{ V_{S_{\beta(c)}} - V_{\tilde{S}_c} \right\} + c^2$$
$$\le cE\left(\int_{\tilde{S}_c}^{\infty} \left(\hat{R}_s^{(\tau)} - \hat{R}_s \right)^2 ds \right)$$
$$\qquad + cE\left(\int_{\tilde{S}_c \vee \tau}^{\infty} \left[v(W_s - W_\tau, s - \tau) - \frac{1}{s - \tau + 1} \right] ds \right)$$
$$\qquad + cE\left(\log \frac{(S_{\beta(c)} - \tau)^+ + 1}{(\tilde{S}_c - \tau)^+ + 1} \right) + c^2 .$$

Proposition 4 now yields

$$\lim_{c \to 0} E \left(\int_{\tilde{S}_c \vee \tau}^{\infty} \left[v(W_s - W_\tau, s - \tau) - \frac{1}{s - \tau + 1} \right] ds \right) = 0.$$

Lemma 5 provides

$$\lim_{c \to 0} E \left(\log \frac{(S_{\beta(c)} - \tau)^+ + 1}{(\tilde{S}_c - \tau)^+ + 1} \right) = 0.$$

Appendix

Here we study the approximation of the integrated posterior variance by the one of a standard normal prior. This leads to the proof of Proposition 4. Let $\alpha \in (0, 2)$. Then

$$\int_\tau^{+\infty} \left| v(W_s - W_\tau, s - \tau) - \frac{1}{s - \tau + 1} \right| ds \tag{10}$$

$$\leq \left(\sup_{\tau \leq s < \infty} \left[v(W_s - W_\tau, s - \tau) - \frac{1}{s - \tau + 1} \right]^{1-\alpha} \right)$$

$$\cdot \int_\tau^{+\infty} \left| v(W_s - \tau, s - \tau) - \frac{1}{s - \tau + 1} \right|^\alpha ds.$$

The Hölder inequality with $p = 2/(2 - \alpha)$ and $q = 2/\alpha$ yields

$$\int_\tau^{+\infty} \left| v(W_s - W_\tau, s - \tau) - \frac{1}{s - \tau + 1} \right|^\alpha ds \tag{11}$$

$$\leq \left(\int_\tau^{+\infty} (s - \tau + 1)^{-2\alpha/(2-\alpha)} \right)^{(2-\alpha)/2}$$

$$\cdot \left(\int_0^{+\infty} \left[(s + 1)v(W_s - W_\tau, s - \tau) - 1 \right]^2 ds \right)^{\frac{\alpha}{2}}.$$

In order to bound the second term on the right-hand side in (11), we first derive stochastic differential equations for $\widehat{R}_t^{(\tau)}$ and $[(t - \tau)^+ + 1]\widehat{R}_t^{(\tau)} - (W_t - W_{t \wedge \tau})$. We have

$$\partial_x \mu(x, t) = v(x, t)$$

and

$$\frac{1}{2}\partial_x\partial_x\mu(x, t) + \partial_t\mu(x, t) = -\mu(x, t)v(x, t).$$

Therefore Itô's formula implies (see [6])

$$d\widehat{R}_t^{(\tau)} = d\left(\mu\left(W_t - W_\tau, t - \tau \right) 1_{\{\tau \leq t\}} \right) = \left[v\left(W_t - W_{t\wedge\tau}, (t - \tau)^+ \right) \right] d\overline{W}_t^{(\tau)}$$

and further

$$d\left([(t - \tau)^+ + 1]\widehat{R}_t^{(\tau)} - (W_t - W_{t\wedge\tau}) \right) \tag{12}$$

$$= \left[\left((t - \tau)^+ + 1 \right)v\left(W_t - W_{t\wedge\tau}, (t - \tau)^+ \right) - 1 \right] d\overline{W}_t^{(\tau)}.$$

We will now prove an alternative representation of $[(t - \tau)^+ + 1]\widehat{R}_t^{(\tau)} - (W_t - W_{t\wedge\tau})$. We first note two useful facts. Since $E|H(Y)\log|H(Y)|| < \infty$, Doob's inequality provides

$$E\left[\sup_{0 \leq t < \infty} E(H(Y)|\mathcal{F}_t^{W,\tau}) \right] < \infty. \tag{13}$$

Moreover $E(Y^2) < \infty$ implies

$$E\left[\sup_{0 \le t < \infty} E(Y|\mathcal{F}_t^{W,\tau})\right] < \infty. \tag{14}$$

Lemma 6 *If G satisfies (A1) to (A3), then*

$$E\left(H(Y)|\mathcal{F}_t^{W,\tau}\right)1_{\{\tau \le t\}} = (t-\tau)^+\widehat{R}_t^{(\tau)} - (W_t - W_{t \wedge \tau}). \tag{15}$$

In particular $[(t-\tau)^+ + 1]\widehat{R}_t^{(\tau)} - (W_t - W_{t \wedge \tau})$ is a martingale with respect to the filtration $\mathcal{F}_t^{W,\tau}$ under P for $\tau \le t < \infty$. Moreover

$$E\left[\sup_{0 \le t < \infty} \left|[(t-\tau)^+ + 1]\widehat{R}_t^{(\tau)} - (W_t - W_{t \wedge \tau})\right|\right] < \infty \tag{16}$$

and thus

$$E\left[\int_\tau^\infty \left((s-\tau+1)v(W_s - W_\tau, s - \tau) - 1\right)^2 ds\right]^{1/2} < \infty. \tag{17}$$

Proof. Let

$$f(x,t) = \int_{-\infty}^{+\infty} e^{yx - \frac{y^2}{2}t}g(y)dy.$$

We have

$$(t-\tau)^+\widehat{R}_t^{(\tau)} - (W_t - W_{t \wedge \tau})$$
$$= E\left((t-\tau)^+Y - (W_t - W_{t \wedge \tau})|\mathcal{F}_t^{W,\tau}\right)$$
$$= \frac{\int_{-\infty}^{+\infty}\left((t-\tau)y - (W_t - W_\tau)\right)e^{y(W_t - W_\tau) - \frac{y^2}{2}(t-\tau)}g(y)dy}{f(W_t - W_\tau, t - \tau)}1_{\{\tau \le t\}}.$$

Lemma 1 of [8] yields for all $B > 0$ and $A \in (-\infty, +\infty)$

$$\int_{-\infty}^{+\infty}(By - A)e^{Ay - B\frac{y^2}{2}}g(y)dy = \int_{-\infty}^{+\infty}\left[\frac{g'(y)}{g(y)}\right]e^{Ay - B\frac{y^2}{2}}g(y)dy.$$

This provides

$$(t-\tau)^+\widehat{R}_t^{(\tau)} - (W_t - W_{t \wedge \tau}) = \frac{\int_{-\infty}^{+\infty} H(y)e^{y(W_t - W_\tau) - \frac{y^2}{2}(t-\tau)}g(y)dy}{f(W_t - W_\tau, t - \tau)}1_{\{\tau \le t\}}$$
$$= E\left(H(Y)|\mathcal{F}_t^{W,\tau}\right)1_{\{\tau \le t\}}.$$

Representation (12), (13), (14), and the Davis–inequality now yield (17).

The proof of Proposition 1 follows now by (7) with $\alpha = 1$ and by (14).

References

1. Beibel, M. (1996). A note on Ritov's Bayes approach to the minimax property of the cusum procedure. Ann. Stat. **24**, 1804–1812
2. Beibel, M. (1997). Sequential change-point detection in continuous time when the post-change drift is unknown. Bernoulli **3**, 457–478
3. Beibel, M. (2000). Sequential detection of signals with known shape and unknown magnitude. Stat. Sinica **10**, 715–729
4. Lerche, H.R. (1986). The shape of Bayes tests of power one. Ann. Stat. **14**, 1030–1048
5. Liptser, R.S., Shiryaev, A.N. (1977). *Statistics of Random Processes* **1**. Springer-Verlag, Berlin
6. Paulsen, V. (1999). A martingale approach for detecting the drift of a Wiener process. Stoch. Proc. Appl. **80**, 177–191
7. Ritov, Y. (1990). Decision theoretic optimality of the CUSUM procedure. Ann. Stat. **18**, 1464–1469
8. Stein, C. (1986). Approximate Computation of Expectations. *IMS Lecture Notes-Monograph Series* **7**, IMS, Hayward, CA
9. Wong, E., Hajek, B. (1984). *Stochastic Processes in Engineering Systems*. Springer-Verlag, Berlin

Box–Cox Transformation for Semiparametric Comparison of Two Samples

Konstantinos Fokianos

Department of Mathematics and Statistics, University of Cyprus, Nicosia, Cyprus

Abstract. We consider the density ratio model which specifies a linear parametric function of the log–likelihood ratio of two densities without assuming any specific form about them and has been found useful for semiparametric comparison of two samples. We study the Box–Cox family of transformations in the context of the density ratio model to suggest a data driven method for identification of the model's true parametric part. The methodology is illustrated by a real data example.

1 Introduction

Quite often in applications we come across with the problem of comparing two samples. The parametric theory resolves the question by appealing to the well known t-test. Accordingly, if $\{X_1, \ldots, X_{n_0}\}$ and $\{X_{n_0+1}, \ldots, X_n\}$ are two *independent* samples with $\bar{X}_0 = \sum_{i=1}^{n_0} X_i/n_0$ and $\bar{X}_1 = \sum_{i=n_0+1}^{n} X_i/n_1$ denoting their respective sample means, then it is well known that the two sample t-test rejects the hypothesis of means equality when

$$\frac{\bar{X}_0 - \bar{X}_1}{S\sqrt{\frac{1}{n_0} + \frac{1}{n_1}}} \geq c \tag{1}$$

where

$$S^2 = \frac{\sum_{i=1}^{n_0} \left(X_i - \bar{X}_0\right)^2 + \sum_{i=n_0+1}^{n} \left(X_i - \bar{X}_1\right)^2}{n - 2},$$

and $n_1 = n - n_0$. The critical value c is determined by the t distribution with $n-2$ degrees of freedom. To carry out test (1), both samples are assumed to be normally distributed with common unknown variance and unknown means. The two sample t-test enjoys several optimality properties, for instance it is an uniformly most powerful unbiased test (see [4]).

Occasionally some (or all) of the needed assumptions fail so that (1) cannot be applied directly. A case in point is illustrated by Fig. 1(a) which displays boxplots of rainfall amounts from two groups of clouds. One group has been seeded with silver nitrate while the other has not. There is a total of 26 observations in each group and the purpose of the experiment was to determine whether cloud seeding increases rainfall. The data are available at http://lib.stat.cmu.edu/DASL/Stories/CloudSeeding.html.

Figure 1(a) shows that both groups follow skewed distributions with large positive values. Clearly both assumptions of normality and equality of variances fail and therefore application of the two sample t-test is question-able. The problem may be bypassed after a logarithmic transformation which leads to symmetric distributions for both groups of clouds with approximately equal variances–see Fig. 1(b).

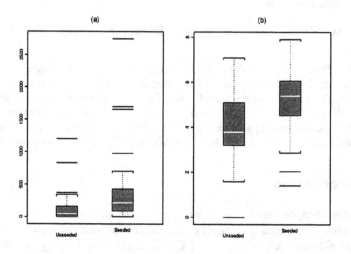

Fig. 1. (a) Boxplots of the clouds data. (b) Boxplots of the clouds data after log transformation.

If a logarithmic (or any other) transformation is not desirable, then we can appeal to the nonparametric theory which approaches the problem of comparing two samples by the so called Mann–Whitney–Wilcoxon test

$$W = \sum_{i=n_0+1}^{n} R_i, \tag{2}$$

where R_i denotes the rank of $\{X_{n_0+1}, \ldots, X_n\}$ among all n observations. For instance, the hypothesis of no shift between the two samples against the alternative of positive shift is rejected for large values of W. For further discussion on test (2), see [7].

Here we consider a quite different approach to the two samples comparison problem. The methodology is relatively new and appeals on the so called *density ratio model* for *semiparametric* comparison of two samples. To be more specific assume that

$$X_1, \ldots, X_{n_0} \sim f_0(x)$$
$$X_{n_0+1}, \ldots, X_n \sim f_1(x) = \exp\left(\alpha + \beta h(x)\right) f_0(x). \tag{3}$$

where $f_i(x)$, $i = 0, 1$ are probability densities, h is a *known* function and α, β are two unknown parameters. In principle, $h(x)$ can be *multivariate* but we assume for simplicity that it is an univariate function throughout the presentation.

We refer to (3) as the density ratio model since it specifies a parametric function of the log likelihood ratio of two densities without assuming any specific form about them. Hence it is a semiparametric model and it is easy to see that under the hypothesis $\beta = 0$, both of the distributions are identical. Consequently if $\hat{\beta}$ stands for the maximum likelihood estimator of β (see (8)) then the following test procedure

$$Z = \frac{\hat{\beta}}{\sqrt{\widehat{\mathrm{Var}(\hat{\beta})}}} \tag{4}$$

where $\widehat{\mathrm{Var}(\hat{\beta})}$ denotes the estimated variance of $\hat{\beta}$, rejects the hypothesis $\beta = 0$ when $\mid Z \mid > c^\star$. The critical value $c^\star$ is determined by the standard normal distribution. Recent contributions on semiparametric inference about the density ratio model include [5],[6], and more recently [3].

2   The Density Ratio Model

To motivate (3) consider the logistic model which has been widely used in applications for the analysis of binary data (see [1] for example). Suppose that Y is a binary response variable and let X be a covariate. The simple logistic regression model is of the form

$$P[Y = 1 \mid X] = \frac{\exp(\alpha^\star + \beta h(x))}{1 + \exp(\alpha^\star + \beta h(x))}, \tag{5}$$

where $\alpha^\star$ and β are scalar parameters. Notice that the marginal distribution of X is left completely unspecified. Assume that $X_1, \ldots, X_{n_0}$ is a random sample from $F(x \mid Y = 0)$. Independent of the X_i, assume that $X_{n_0+1}, \ldots, X_n$ is a random sample from $F(x \mid Y = 1)$, and let $n_1 = n - n_0$. Put $\pi = P(Y = 1) = 1 - P(Y = 0)$ and assume that $f(x \mid Y = i) = dF(x \mid Y = i)/dx$ exists and represents the conditional density function of X given $Y = i$ for $i = 0, 1$. A straightforward application of the Bayes theorem shows that

$$\frac{f(x \mid Y = 1)}{f(x \mid Y = 0)} = \exp(\alpha + \beta h(x))$$

with $\alpha = \alpha^\star + \log[(1 - \pi)/\pi]$. In other words, model (5) is equivalent to the following two sample semiparametric model

$$X_1, \ldots, X_{n_0} \sim f(x \mid Y = 0)$$
$$X_{n_0+1}, \ldots, X_n \sim f(x \mid Y = 1) = \exp(\alpha + \beta h(x)) f(x \mid Y = 0),$$

with $\alpha = \alpha^\star + \log[(1 - \pi)/\pi]$–a fact that leads to (3).

2.1   An Example

To show that (3) and (5) are not meaningless, let Y be a binary random variable with $P(Y = 1) = \pi$. Suppose that given $Y = 0$, X follows the lognormal distribution with parameters μ_0 and σ^2, and given $Y = 1$, X follows the lognormal distribution with parameters μ_1 and σ^2. Then, the ratio of conditional densities is exponential:

$$\frac{f(x|Y = 1)}{f(x|Y = 0)} = \exp(\alpha + \beta \log x)$$

with

$$\alpha = \frac{\mu_0^2 - \mu_1^2}{2\sigma^2}, \quad \beta = \frac{\mu_1 - \mu_0}{\sigma^2}, \quad h(x) = \log x. \tag{6}$$

This implies the logistic regression model (5) because upon defining α^* by the equation $\alpha^* = \log[(1 - \pi)/\pi] + \alpha$, Bayes rule gives

$$\frac{P(Y = 1 \mid X)}{P(Y = 0 \mid X)} = \exp(\alpha^* + \beta \log x)$$

and since $P(Y = 0 \mid X) = 1 - P(Y = 1 \mid X)$ we obtain (5).

2.2   Box–Cox Transformation for the Density Ratio Model

Recall (3) and assume that the data are positive, that is all $X > 0$. The density ratio model (3) depends clearly on the choice of the function h which needs to be known. To relax this assumption, we assume that h is parameterized according the so called Box–Cox family of transformations (see [2])

$$h_\lambda(x) = \begin{cases} \dfrac{x^\lambda - 1}{\lambda} & \text{when } \lambda \neq 0 \\ \log x & \text{when } \lambda = 0. \end{cases}$$

Thus expression (3) becomes

$$X_1, ..., X_{n_0} \sim f_0(x)$$
$$X_{n_0+1}, ..., X_n \sim f_1(x) = \exp(\alpha + \beta h_\lambda(x)) f_0(x). \tag{7}$$

It turns out that the Box–Cox family of transformations enlarges the density ratio model by providing a data driven choice of $h(x)$. In this respect the data analyst can identify the appropriate $h(x)$ in applications. The following section discusses inference regarding model (7). Extensions for multivariate $h(x)$ are briefly sketched in Section 4.

3   Inference

Inference can be carried out along the lines of [6]. Accordingly, it can be shown that inference for model (7) is based on the following empirical log likelihood

$$l(\alpha, \beta, \lambda) = -\sum_{i=1}^{n} \log\left[1 + \rho_1 \exp\left(\alpha + \beta h_\lambda(x_i)\right)\right] + \sum_{i=n_0+1}^{n} \left(\alpha + \beta h_\lambda(x_i)\right), \quad (8)$$

with $\rho_1 = n_1/n_0$. Expression (8) has been derived after profiling out an infinite dimensional parameter, namely the cumulative distribution function of $f_0(x)$, say $F_0(x)$. The key concept is that of the empirical likelihood (see [8]).

Estimation of λ proceeds by following a standard procedure. To be more specific we maximize equation (8) for given λ with respect to α and β. If we denote by $l_{\max}(\lambda)$ the maximized log likelihood for a given value of λ, then a plot of $l_{\max}(\lambda)$ against λ for a trial series of values will reveal $\hat{\lambda}$–the maximum likelihood estimator of λ. An approximate $100(1 - a)\%$ confidence interval for λ consists of those values of λ which satisfy the inequality

$$l_{\max}(\hat{\lambda}) - l_{\max}(\lambda) \leq \frac{1}{2}\chi_{1;1-a}^2 \quad (9)$$

where $\chi_{1;1-a}^2$ is the percentage point of the chi–squared distribution with one degree of freedom which leaves an area of a in the upper tail of the distribution.

3.1   Application

Figure 2 illustrates the above methodology applied to clouds data. In other words this is a plot of the maximized log likelihood as λ varies in $[-2, 2]$ with step equal to 0.01. The maximum value is obtained at $\hat{\lambda} = 0.18$. The horizontal line indicates a 90% confidence interval for λ–according to (9)–which turns out to be $[-0.58, 1.50]$. Consequently, values of λ equal to -1/2, 0, 1/2, 1 and 3/2 are not excluded as possibilities by the data. Apparently the relative small number of observations lead to negligible changes to the log likelihood for different λ and therefore the obtained confidence interval is rather large. Hence it is preferable to use values that fall near the viscinity of the maximum. For the clouds data we choose $\lambda = 0, 1/2$. This discussion confirms from another point of view that log transformation is appropriate for the data at hand.

Table 1 lists the testing results for the clouds data. That is, the first two rows report the results of the ordinary t–test (1) for both raw and log transformed data, and the third row refers to the output of the Mann–Whitney–Wilcoxon test (2). The last two rows of Table 1 list the values of the test

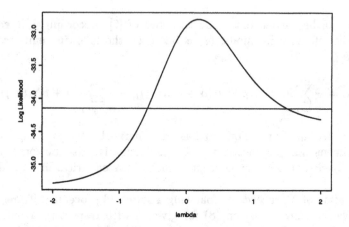

Fig. 2. Values of the log likelihood for the clouds data when λ varies in $[-2, 2]$. The horizontal line indicates a 90% confidence interval for λ.

statistic (4) for $h(x) = \log(x)$ ($\lambda=0$) and $h(x) = \sqrt{x}$ ($\lambda=1/2$), respectively. Notice that all the test procedures–besides the ordinary t–test–reject the hypothesis of identical population. In addition both t, after log transformation of the data, and Wilcoxon tests support the rejection slightly stronger than (4) for $\lambda = 0$. To conclude the example we point out that the estimator of β is equal to 0.450 (0.036 respectively) with an estimated standard error 0.192 (0.016 respectively) for $\lambda = 0$ ($\lambda = 0.5$ respectively). The positive sign of $\hat{\beta}$ in both cases indicates that the population of seeded clouds assumes larger values than the population of unseeded clouds. This is an indication that cloud seeding increases rainfall.

Table 1. Testing for the clouds data

Procedure	Test Statistic	p–value
t-test	-1.998	0.0511
t-test after log transformation	-2.544	0.0141
Wilcoxon	-2.461	0.0138
Z with $\lambda = 0$	2.343	0.0191
Z with $\lambda = 0.5$	2.250	0.0244

3.2   Further Results

If it is desirable to calculate the exact value of $\hat{\lambda}$, then maximization of (8) with respect to all the parameters leads to the maximum empirical likelihood estimator of the parameter vector $(\alpha, \beta, \lambda)'$ –say $(\hat{\alpha}, \hat{\beta}, \hat{\lambda})'$. Under suitable regularity condition, the maximum likelihood estimator of the parameter vector $(\alpha, \beta, \lambda)'$ is asymptotically normally distributed. That is

$$\sqrt{n} \begin{pmatrix} \hat{\alpha} - \alpha \\ \hat{\beta} - \beta \\ \hat{\lambda} - \lambda \end{pmatrix} \to \text{Normal}(0, \Sigma)$$

as $n \to \infty$ where Σ is the asymptotic covariance matrix. An empirical assessment of this fact is manifested by Fig. 3 which displays Q–Q plots of estimators based on 100 simulations with $n_0 = 300$ and $n_1 = 500$. Those are derived by assuming that $f_0(x)$ is lognormal with parameters 1 and 1 and $f_1(x)$ is lognormal with parameters 4 and 1. Obviously, equation (6) yields $\alpha = -7.5$, $\beta = 3$ and $\lambda = 0$. The Q-Q plots indicate that the asymptotic normality is valid at least for large sample sizes.

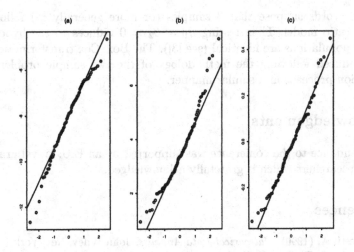

Fig. 3. Q–Q plots of estimators based on 100 simulations assuming $f_0(x)$ is lognormal with parameters 1 and 1 and $f_1(x)$ is lognormal with parameters 4 and 1 with $n_0 = 300$ and $n_1 = 500$. (a) $\hat{\alpha}$ (b) $\hat{\beta}$ (c) $\hat{\lambda}$.

4   Extensions

4.1   Multivariate $h(x)$

Consider the following situation

$$X_1, ..., X_{n_0} \sim f_0(x)$$
$$X_{n_0+1}, ..., X_n \sim f_1(x) = \exp\left(\alpha + \beta' h(x)\right) f_0(x).$$

where β is a p-dimensional vector of parameters and $h(x)$ is a p–dimensional function. Assuming that the data are positive, then the Box–Cox transformation can be applied to $h(x)$ componentwise for identification of the true functional form of the model. However the issues of estimation and testing become more complicated.

4.2   Semiparametric ANOVA

Model (3) can be generalized by considering more than two samples. Indeed, suppose that

$$X_{01}, ..., X_{0n_0} \sim f_0(x)$$
$$X_{11}, ..., X_{1n_1} \sim f_1(x) = \exp\left(\alpha_1 + \beta_1 h(x)\right) f_0(x)$$
$$X_{21}, ..., X_{2n_2} \sim f_2(x) = \exp\left(\alpha_2 + \beta_2 h(x)\right) f_0(x).$$

In other words suppose that 3 samples (or more generally m) follow the density ratio model. Then testing $\beta_1 = \beta_2 = 0$ reduces to the hypothesis that all populations are identical (see [3]). The Box–Cox transformation can be introduced following the methodology of the two sample problem and estimation proceeds in a similar manner.

Acknowledgements

My attendance to the conference was supported by an E.U. travel grant for young researchers which is gratefully acknowledged.

References

1. Agresti, A. (1990). *Categorical Data Analysis*. John Wiley, New York.
2. Box, G. E. P., Cox, D. R. (1964). An Analysis of Transformations (with discussion). J. Roy. Stat. Soc. B. **26**, 211–252
3. Fokianos, K., Kedem, B., Qin, J., Short, D. (2001). A Semiparametric Approach to the One–Way Layout. Technometrics **43**, 56–65
4. Lehmann E. L. (1997). *Testing Statistical hypotheses* 2nd edition. Springer, New York

5. Qin, J. (1998). Inference for Case–Control and Semiparametric Two–Sample Density Ratio Model. Biometrika **85**, 619–630

6. Qin, J., Zhang, B. (1997). A Goodness of Fit Test for Logistic Regression Model Based on Case–Control Data. Biometrika **84**, 609–618

7. Randles, R. H., Wolfe, D. A. (1979). *Introduction to the Theory of Nonparametric Statistics*. John Wiley, New York

8. Owen, A. B. (1988). Empirical Likelihood Ratio Confidence Intervals for a Single Functional. Biometrika **75**, 237–249

Minimax Nonparametric Goodness-of-Fit Testing

Yuri I. Ingster[1]* and Irina A. Suslina[2]

[1] St.Petersburg State Transport University, Russia
[2] St.Petersburg Institute for Exact Mechanics and Optics, Russia

Abstract. We discuss and study minimax nonparametric goodness-of-fit testing problems under Gaussian models in the sequence space and in the functional space. The unknown signal is assumed to vanish under the null-hypothesis. We consider alternatives under two-side constraints determined by Besov norms. We present the description of the types of sharp asymptotics under the sequence space model and of the rate asymptotics under the functional model. The structures of asymptotically minimax and minimax consistent test procedures are given. These results extend recent results of the paper [12]. The results for an adaptive setting are presented as well.

1 Minimax Setting in Goodness-of-Fit Testing

Let the unknown signal s be observed overlapped with white Gaussian noise. This is the classical *Gaussian functional model* which corresponds to the random process

$$dX_\varepsilon(t) = s(t)dt + \varepsilon dW(t), \ t \in [0,1], \ s \in L_2(0,1) \tag{1}$$

where W is the standard Wiener process and $\varepsilon > 0$ is the noise level. We want to detect a signal, that is, to test the null-hypothesis $H_0 : s = 0$. Certainly the model 1 is equivalent to *the Gaussian sequence model*: we observe an infinite-dimensional random vector x with unknown mean vector θ

$$x = \theta + \varepsilon\xi, \ \theta \in \ell^2, \ \xi = \{\xi_i\}, \ \xi_i \sim \mathcal{N}(0,1), \ \xi_i \text{ are i.i.d.} \tag{2}$$

To obtain the equivalence it suffices to fix an orthonormal basis $\{\phi_i\}$ in $L_2(0,1)$ and consider the Fourier transform $\xi_i = \int_0^1 \phi_i(t)dX_\varepsilon(t)$, $\theta_i = (s,\phi_i)$. Under the model 2 we test the null-hypothesis $H_0 : \theta = 0$.

We consider an asymptotic variant of the minimax setting in hypothesis testing. Let us present definitions for the model 1; the translation for the model 2 is evident.

Let alternatives $H_1 : s \in S_\varepsilon$ be given; here $S_\varepsilon \subset L_2(0,1)$. For a test ψ_ε we consider type I error and maximal type II error:

$$\alpha_\varepsilon(\psi_\varepsilon) = E_{\varepsilon,0}\psi_\varepsilon, \ \beta_\varepsilon(\psi_\varepsilon) = \sup_{s \in S_\varepsilon} E_{\varepsilon,s}(1 - \psi_\varepsilon);$$

---

* Research was partially supported by RFFI Grants No 99-01-00111, No 00-15-96019 and by RFFI - DFG (Russia - Germany) Grant No 98-01-04108

here $E_{\varepsilon,s}$ is the expectation over the measure $P_{\varepsilon,s}$ which corresponds to observation 1. Let $\gamma_\varepsilon(\psi_\varepsilon) = \alpha_\varepsilon(\psi_\varepsilon) + \beta_\varepsilon(\psi_\varepsilon)$ be the total error for the test ψ_ε. Denote by γ_ε the *minimal total error* $\gamma_\varepsilon = \inf \gamma_\varepsilon(\psi_\varepsilon)$ where the infimum is taken over all possible tests ψ_ε. Clearly, $0 \le \gamma_\varepsilon \le 1$ *and $\gamma_\varepsilon = \gamma_\varepsilon(S_\varepsilon)$ depends on the set S_ε. We call a minimax hypothesis testing problem *trivial*, if $\gamma_\varepsilon = 1$ for $\varepsilon < \varepsilon_0$ small enough. For the non-trivial case the *sharp asymptotics problem* is to study the asymptotics of minimax total errors γ_ε as $\varepsilon \to 0$ up to vanishing terms and to construct asymptotically minimax tests ψ_ε such that $\gamma_\varepsilon(\psi_\varepsilon) = \gamma_\varepsilon + o(1)$ as $\varepsilon \to 0$. The *rate asymptotic problem* is to study conditions on the sets S_ε either for $\gamma_\varepsilon \to 1$ or for $\gamma_\varepsilon \to 0$; in the last case we would like to construct *minimax consistent tests* ψ_ε such that $\gamma_\varepsilon(\psi_\varepsilon) \to 0$ as $\varepsilon \to 0$.

There are some difficulties to use the minimax approach in goodness-of-fit testing problem where only the null-hypothesis is given. The natural alternative $H_1 : s \ne 0$ corresponds to the set $S_\varepsilon = L_2(0,1) \setminus \{0\}$. However the minimax problem is trivial in this case because the alternative contains points s "arbitrary close" to the point 0. To overcome this difficulty we need to remove some small neighborhood of the null-hypothesis, that is, to consider an one-side constraint $\|s\|_{(1)} \ge \rho_\varepsilon$ where $\|\cdot\|_{(1)}$ is some norm defined on the linear space $L_2(0,1)$ (this may be no L_2-norm) and ρ_ε is a family of radii. However for main norms of interest the problem is trivial as well. In particular, this holds for L_p-norms with $1 \le p \le \infty$ and for all $\varepsilon > 0, \rho_\varepsilon > 0$. This fact was shown by [6] for $p = 2$ and this follows from [1] for $p \ne 2$; see [10]. The reason is that there are not any regularity conditions for unknown signals.

For this reason we need to add some regularity constraint. It is convenient to characterize regularity of a signal by some other norm $\|\cdot\|_{(2)}$. This leads to *alternatives under two-side constraints*:

$$S_\varepsilon = \{s \in L_2(0,1) : \quad \|s\|_{(1)} \ge \rho_\varepsilon, \ \|s\|_{(2)} \le R\}. \tag{3}$$

Analogously, under the model 2 we consider the alternatives

$$\Theta_\varepsilon = \{\theta \in \ell^2 : \quad |\theta|_{(1)} \ge \rho_\varepsilon, \ |\theta|_{(2)} \le R\} \tag{4}$$

where $|\cdot|_{(1)}, |\cdot|_{(2)}$ are some norms in the sequence space ℓ^2.

2   Comparing with the Minimax Estimation Problem

Thus, we have two norms which determine the problem. Analogous norms are used in the *minimax estimation problem*. Namely, in the estimation problem the quality of an estimator $s_\varepsilon = s_\varepsilon(X_\varepsilon)$ for a signal s is characterized by a risk function $R_\varepsilon(s, s_\varepsilon) = E_{\varepsilon,s} l(s, s_\varepsilon)$. Here l is a loss-function which typically

---

* To obtain the right-hand side inequality it suffices to consider the trivial test $\psi_\varepsilon(X_\varepsilon) = \alpha, \ \alpha \in [0,1]$.

is of the form $l(s, s_\varepsilon) = F(\|s - s_\varepsilon\|_{(1)})$; this is determined by a non-decreasing function $F(t)$, $t \geq 0$; $F(0) = 0$ and by a distance between the signal s and the estimator s_ε in some norm $\| \cdot \|_{(1)}$. Minimax quality of an estimator s_ε is characterized by maximal risk $R_\varepsilon(s_\varepsilon, S) = \sup_{s \in S} R_\varepsilon(s, s_\varepsilon)$ where $S \subset L_2(0,1)$ is a given set of unknown signals. The object of the study is the asymptotics of minimax risk $R_\varepsilon(S) = \inf_{s_\varepsilon} R_\varepsilon(s_\varepsilon, S)$ where the infimum is taken over all possible estimators. If $S = L_2(0,1)$, then for the main losses of interest it is impossible to construct minimax consistent estimators such that $R_\varepsilon(s_\varepsilon, S) \to 0$ (see [7]). For this reason one adds a regularity constraint which typically is of the form

$$S = \{s \in L_2(0,1) : \quad \|s\|_{(2)} \leq R\};$$

here the norm $\| \cdot \|_{(2)}$ characterizes a regularity of a signal.

Usually one determines losses either by L_2-norm or by L_p-norms. However sometimes one is interested in estimation not only a signal, but its derivatives of degree σ. These corresponds to the losses determined by the Sobolev norm $\| \cdot \|_{\sigma,p}$ or by the Besov norm $\| \cdot \|_{\sigma,p,h}$ (see [22] for definitions of the norms, for its properties and embedding theorems). Usually a regularity is characterized by a Sobolev norm $\| \cdot \|_{\eta,q}$ or by a Besov norm $\| \cdot \|_{\eta,q,t}$ with some other parameters, that is, one has

$$\| \cdot \|_{(1)} = \| \cdot \|_{\sigma,p} \text{ or } \| \cdot \|_{(1)} = \| \cdot \|_{\sigma,p,h}; \quad \| \cdot \|_{(2)} = \| \cdot \|_{\eta,q} \text{ or } \| \cdot \|_{(2)} = \| \cdot \|_{\eta,q,t}. \quad (5)$$

The minimax signal estimation problem and related estimation problems under the regression function model and the probability density model were studied very intensively. The main problems of interest were the study of rates of the minimax risk $R_\varepsilon(S)$ and the construction of estimators s_ε such that $R_\varepsilon(S) \sim R_\varepsilon(s_\varepsilon, S)$. Typically the rates are of the form $R_\varepsilon(S) = F(\hat{\rho}_\varepsilon)$; here F is the function which determines losses and the quantities $\hat{\rho}_\varepsilon = \hat{\rho}_\varepsilon(S)$ characterizes the rates of possible accuracy of estimation in the norm $\| \cdot \|_{(1)}$ for the set S. Typically the rates $\hat{\rho}_\varepsilon$ are essentially larger than classical parametrical rates $\hat{\rho}_\varepsilon = \varepsilon$ and these depend on the norm $\| \cdot \|_{(1)}$ and on the norm $\| \cdot \|_{(2)}$ which determines the set S. For the norms 5 with $\sigma \geq 0$, $\eta - \sigma > \min(p^{-1} - q^{-1}, 0)$, the estimation rates $\hat{\rho}_\varepsilon$ are of the form

$$\hat{\rho}_\varepsilon \sim \begin{cases} \varepsilon^{2C}, & \text{if } \eta/p - \sigma/q > 1/2q - 1/2p, \\ (\varepsilon^2 \log(\varepsilon^{-1})^D, & \text{if } \eta/p - \sigma/q < 1/2q - 1/2p, \end{cases}$$

where

$$C = \frac{\eta - \sigma}{2\eta + 1}, \qquad D = \frac{\eta - \sigma - q^{-1} + p^{-1}}{2\eta - 2q^{-1} + 1}.$$

Sharp asymptotics in minimax estimation have been studied for special cases only. Often estimators s_ε, which provide optimal rates, depend on parameters of the norms $\kappa = (\sigma, \eta, p, q)$ in the problem, which often are not specified in problems of interest. This leads to an adaptive setting: to construct estimators which provide optimal or nearly optimal rates for a wide range of parameters κ. Adaptive estimation problems have been studied as well. It was shown that often it is impossible to construct adaptive estimators without some losses in the rates; however these loses are small enough (one needs to add log factor to the rates). See papers [2], [4], [15], [16], [13], [14], [19] and references in these papers.

One of the main methods for the study of minimax estimation problems has been proposed by Donoho and Johnstone. This is based on the *wavelet transform* of the model 2 which corresponds to a regular wavelet basis $\{\phi_{ij}\}$. The problem is studied in the sequence space ℓ^2 of pyramidal structure: $\theta = \{\theta_{ij}, j > 0, i = 1, ..., 2^j\}$ with Besov norms in the sequence space:

$$\|\cdot\|_{(1)} = |\cdot|_{r,p,h}, \quad \|\cdot\|_{(2)} = |\cdot|_{s,q,t}; \quad r = \sigma + 1/2 - 1/p, \quad s = \eta + 1/2 - 1/q \quad (6)$$

which are defined by the equality

$$|\theta|_{r,p,h} = \left(\sum_j 2^{jrh} \left(\sum_i |\theta_{ij}|^p\right)^{h/p}\right)^{1/h}; \quad 0 < p, h < \infty; \quad -\infty < r < \infty$$

with a simple modification for $p, h = \infty$; the norms $|\theta|_{s,q,t}$ are analogous. The Besov norms in the sequence space 6 are equivalent to the Besov norms in the functional space 5 (at least for $\sigma \geq 0$, $\eta \geq 0$, $p \geq 1$, $q \geq 1$ and up to a finite-dimensional subspace** ; see [4], [5]).

3 Types of Asymptotics in Goodness-of-Fit Testing

By natural analogy between estimation and hypothesis testing problems under the functional Gaussian noise model 1 we study the alternatives determined by the norms 5 with $\sigma, \eta \geq 0$, $1 \leq p, q, h, t \leq \infty$. Under the model 2 we consider the alternatives 4 for the norms 6; however we study not only classical norms, but quasi-norms with $-\infty < \sigma, \eta < \infty$, $0 < p, q, h, t \leq \infty$ as well. Note that a negative parameter $\sigma < 0$ corresponds to the estimation of the integral of the a signal; in particular under the probability density model the case $\eta = -1$ corresponds to estimation of the distribution function.

We obtain the sharp asymptotics for the norms 6 with $h \leq p$, $t \geq q$ and the rate asymptotics without the last constraints. These lead to the rate asymptotics for the norms 5. Note that analogous problems have been studied before; see [10], [12], [18], [17], [20], [21] for references.

The study is based on the constructions of asymptotically least favorable priors in the problem. The results that we present below are an extension of

---

** It seems that these constraints are not essential for equivalence between Besov norms in functional space and ones in the sequence space under the regular wavelet transform. However we do not know any references on this topic.

the results of the paper [12] where sharp asymptotics are presented for power norms in the sequence space $|\cdot|_{(1)} = |\cdot|_{r,p}$, $|\cdot|_{(2)} = |\cdot|_{s,q}$ of the form

$$|\theta|_{r,p} = \left(\sum_{i=1}^{\infty} i^{rp} |\theta_i|^p \right)^{1/p}. \tag{7}$$

We are basing on methods developed in [12] which provide a reduction of the minimax hypothesis testing problem to a convex extreme problem on a specific Hilbert space. An extension to the adaptive problem is given in [11].

Let us start from the main types of the asymptotics for alternatives of the type 4. (The partitions of the space of parameters into regions of various types of asymptotics is presented in Fig. 1 – 6 below).

Trivial type coresponds to the trivial problem (that is, $\gamma_\varepsilon \equiv 1$). This type arises for $\sigma \geq \sigma_p$ without the second constraint in 4 (that is, for $R = \infty$) where

$$\sigma_p = \begin{cases} -1/4, & p \leq 2 \\ 1/(2p) - 1/2, & p > 2 \end{cases}.$$

Under the second constraint triviality is possible, in particular, for $\eta - \sigma \leq \min(q^{-1} - p^{-1}, 0)$. We denote this type by T.

For non-trivial problems the main questions of interest are: what are the radii ρ_ε to obtain either $\gamma_\varepsilon(\Theta_\varepsilon) \to 0$ or $\gamma_\varepsilon(\Theta_\varepsilon) \to 1$? A typical answer is given in terms of *rates of testing (or critical radii)* ρ_ε^*:

$$\gamma_\varepsilon(\Theta_\varepsilon) \to 0, \quad \text{as } \rho_\varepsilon/\rho_\varepsilon^* \to \infty; \qquad \gamma_\varepsilon(\Theta_\varepsilon) \to 1, \quad \text{as } \rho_\varepsilon/\rho_\varepsilon^* \to 0. \tag{8}$$

If alternatives are simple or of finite dimension, typical rates are classical: this corresponds to $\rho_\varepsilon^* = \varepsilon$. We call this case *classical* and denote this type by C. This corresponds to $\sigma < \sigma_p$; the second constraint is not necessary in this case.

The main types of the asymptotics are *Gaussian*: these correspond to the asymptotics of the type

$$\gamma_\varepsilon(\Theta_\varepsilon) = 2\Phi(-u_\varepsilon/2) + o(1). \tag{9}$$

The quantities $u_\varepsilon = u_\varepsilon(\tau, \rho_\varepsilon, R)$ (here and below $\tau = (\kappa, h, t)$) characterize the quality of testing in the problem; the rates ρ_ε^* are determined by the relation: $u_\varepsilon \sim 1$. Asymptotically least favorable priors are of product type:

$$\pi^\varepsilon(d\theta) = \prod_j \prod_i \pi_{\varepsilon,ij}(d\theta_{ij})$$

where the factors $\pi_{\varepsilon,ij} = \pi(z_{\varepsilon,j}, h_{\varepsilon,j})$ do not depend on i and these are symmetrical three-point measures on the real line:

$$\pi(z, h) = (1 - h)\delta_0 + \frac{h}{2}(\delta_{-z} + \delta_z), \quad h \in [0,1], \; z \geq 0 \tag{10}$$

(or two-points measure for $h = 1$). The sequences $\bar{z}_\varepsilon = \{z_{\varepsilon,j}\}$, $\bar{h}_\varepsilon = \{h_{\varepsilon,j}\}$ are determined as the solution of a specific extreme problem. The log-likelihood ratio for the priors is asymptotically Gaussian in this case. The quantities u_ε are of the form

$$u_\varepsilon^2 = u_\varepsilon^2(\tau, \rho_\varepsilon, R) = \sum_j 2^{j+1} h_{\varepsilon,j}^2 \sinh^2(z_{\varepsilon,j}^2/2) \tag{11}$$

and these depend essentially on parameters κ which determine the constraints on the alternative (the dependence on parameters h, t is not too essential).

We have different types of Gaussian asymptotics (two main types denoted by G_1 and G_2 and "boundary" types denoted by $G_3 - G_5$). The main types correspond to the following relations:

$$G_1 : \quad u_\varepsilon^2 = c_0 m z_0^4, \quad \text{where} \quad \begin{cases} c_1 m^{\sigma+1/2} z_0 = \rho_\varepsilon/\varepsilon, \\ c_2 m^{\eta+1/2} z_0 = R/\varepsilon, \end{cases} \tag{12}$$

$$G_2 : \quad u_\varepsilon^2 = c_0 n h_0^2, \quad \text{where} \quad \begin{cases} c_1 n^{\sigma+1/2} h_0^{1/p} = \rho_\varepsilon/\varepsilon, \\ c_2 n^{\eta+1/2} h_0^{1/q} = R/, \varepsilon \end{cases} \tag{13}$$

The quantities $m = m_\varepsilon \to \infty$, $n = n_\varepsilon \to \infty$ have the sense of "effective dimensions" in the problem. For the type G_1 the main term consists of factors with $h_{\varepsilon,j} = 1$ and typical quantities $z_{\varepsilon,j} \sim z_0 = z_{0\varepsilon}$; on the other hand, for the type G_2 the main term consists of factors with $z_{\varepsilon,j} \sim 1$ and typical quantities $h_{\varepsilon,j} \sim h_0 = h_{0\varepsilon}$. For $h \leq p$, $t \geq q$ the quantities $c_l = c_l(j_\varepsilon, \tau)$, $l = 0, 1, 2$ are positive 1-periodical functions on $j_\varepsilon = \log_2 m_\varepsilon$ for the type G_1 and on $j_\varepsilon = \log_2 n_\varepsilon$ for the type G_2.

Using 12, 13 we have

$$u_\varepsilon^2(\tau, \rho_\varepsilon, R) \sim d\rho_\varepsilon^{A_l} R^{-B_l} \varepsilon^{B_l - A_l} \tag{14}$$

where the constants A_l, B_l correspond to the types G_l, $l = 1, 2$

$$A_1 = \frac{4\eta + 1}{\eta - \sigma}, \qquad\qquad B_1 = \frac{4\sigma + 1}{\eta - \sigma},$$

$$A_2 = \frac{2\eta + 1 - 1/q}{\eta/p - \sigma/q + 1/2p - 1/2q}, \quad B_2 = \frac{2\sigma + 1 - 1/p}{\eta/p - \sigma/q + 1/2p - 1/2q}.$$

The function $d = d(\tau, \rho_\varepsilon, R) = d_1(\tau, x)$ is positive and periodic on $x = \log_2(R/\rho_\varepsilon)$ for the type G_1 and on $x = \log_2(R^q \rho_\varepsilon^{-p} \varepsilon^{p-q})$ for the type G_2. These lead to the rates (we assume R be fixed)

$$\rho_\varepsilon^* = \varepsilon^{c_l}, \quad l = 1, 2; \quad c_1 = \frac{\eta - \sigma}{\eta + 1/4}, \quad c_2 = \frac{2(\eta - \sigma) + p^{-1} - q^{-1}}{2\eta + 1 - q^{-1}}. \tag{15}$$

Frontier types $G_3 - G_5$ correspond to relations analogous to 14 with additional logarithmic factors on h_0, m or n; the degrees of logarithmic factors depend on parameters h, t; the quantities $d = d(\tau)$ do not depend on $\rho_\varepsilon, R, \varepsilon$.

Note that for the region of parameters corresponding to the Gaussian asymptotics the rates in the hypothesis testing problem are usually smaller than the rates in the estimation problem: $\rho_\varepsilon^* = o(\hat{\rho}_\varepsilon)$.

For example, let $p = q = 2$, $\sigma = 0$, $\eta > 0$, that is, we consider η-smooth functions in L_2-norm and removed L_2-ball; this corresponds to the type G_1. Then the rates of estimations are $\hat{\rho}_\varepsilon = \varepsilon^{2\eta/(2\eta+1)}$ and the rates of testing are $\rho_\varepsilon^* = \varepsilon^{4\eta/(4\eta+1)}$.

This means we can distinguish between the null-hypothesis and the alternative which are closer than the rates of the estimation accuracy. Moreover tests based on distance statistics between the null-hypothesis and the best estimators does not provide good rates of testing.

Degenerate asymptotics (we denote this type by D) are possible for

$$\sigma > 1/p - 1/2, \quad \eta - \sigma > 1/q - 1/p, \quad \eta/p - \sigma/q \le 1/2q - 1/2p.$$

For power norms 7 this type is characterized by the relations

$$\gamma_\varepsilon = \Phi(\sqrt{2\log n_\varepsilon} - n_\varepsilon^{-r}\rho_\varepsilon/\varepsilon) + o(1); \tag{16}$$

where the quantities $n_\varepsilon = (R/\rho_\varepsilon)^{1/(s-r)} \to \infty$ have the sense of "efficient dimensions". The asymptotically least favorable priors π_ε are supported on specific orthogonal collections in ℓ^2. The likelihood ratio L_{π_ε} is asymptotically degenerate (that is, $L_{\pi_\varepsilon} = C_\varepsilon + o(1)$ under the null-hypothesis where C_ε are non-random). The rates $\rho_\varepsilon^* = \rho_\varepsilon^*(\kappa, R_\varepsilon)$ are defined by the relation $n_\varepsilon^{-r}\rho_\varepsilon = \varepsilon\sqrt{2\log n_\varepsilon} + O(1)$ which yields

$$\rho_\varepsilon^* = \varepsilon^{(s-r)/s} R^{r/s} \left((2/s)\log(R/\varepsilon)\right)^{(s-r)/2s}. \tag{17}$$

These lead to sharper distinguishability conditions than in 8:

$$\gamma_\varepsilon \to 0, \quad \text{as} \quad \liminf \rho_\varepsilon/\rho_\varepsilon^* > 1; \tag{18}$$

$$\gamma_\varepsilon \to 1, \quad \text{as} \quad \limsup \rho_\varepsilon/\rho_\varepsilon^* < 1. \tag{19}$$

For Besov norms 6 we have analogous relations:

$$\gamma_\varepsilon \to 0, \quad \text{as} \quad \liminf \rho_\varepsilon/\rho_\varepsilon^* > c_1; \tag{20}$$

$$\gamma_\varepsilon \to 1, \quad \text{as} \quad \limsup \rho_\varepsilon/\rho_\varepsilon^* < c_0 \tag{21}$$

with some constants $0 < c_0 \le c_1 < \infty$, Analogous relations hold true under the functional Gaussian model. It was show in the paper [18] that, $c_0 = c_1$ for the case of Sobolev norms in 5 where $\eta \ge 1$ is an integer, $\eta q > 1$ and $\sigma = 0$, $p = \infty$.

If R is a constant, then the rates 17 are the same as in the estimation problem: $\rho_\varepsilon^* \sim \hat{\rho}_\varepsilon$ in the region of degenerate asymptotics.

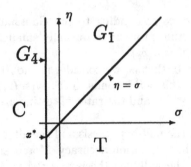

Fig. 1: $p \lesssim 2$, $p < q \leq \infty$

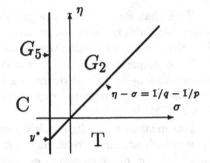

Fig. 2: $2 < p = q < \infty$

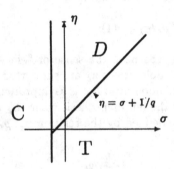

Fig. 3: $q \lesssim p = \infty$

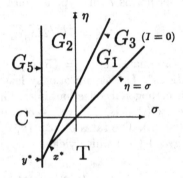

Fig. 4: $2 < p < q \leq \infty$; $p \lesssim 4$

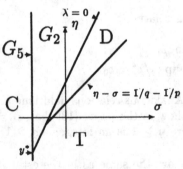

Fig. 5: $p > q$, $p \geq 2$

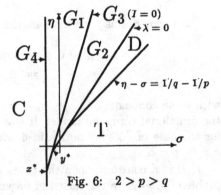

Fig. 6: $2 > p > q$

In Fig. 1–6 we describe the partitions of the plane of parameters $\{\sigma, \eta\}$ into the regions of the asymptotics of different types for fixed values p, q. The partition is determined by the quantities λ, I which are linear functions of σ, η for fixed p, q:

$$\lambda = (\eta + 1/2)/p - (\sigma + 1/2)/q, \quad I = 2(\eta - \sigma) - 4(\eta/p - \sigma/q) + 1/q - 1/p.$$

We denote by $x^* = x^*_{p,q}$ and $y^* = y^*_{p,q}$ the points on the plane $\{\sigma, \eta\}$ with the coordinates $x^* = (-1/4, -1/4)$, $y^* = (1/(2p) - 1/2, 1/(2q) - 1/2)$. Under the Gaussian functional model and for Sobolev and Besov norms we have similar partitions (at least for $\sigma \geq 0$, $\eta \geq 0$, $p \geq 1$, $q \geq 1$).

4   Test Procedures

In the region D asymptotical minimax tests for $pt \leq qh$ (and consistent tests for $pt < qh$, $s/p < r/q$) could be based on the simple threshold procedures:

$$\psi_\varepsilon = 1_{\mathcal{X}^{thr}_\varepsilon}, \qquad \mathcal{X}^{thr}_\varepsilon = \{x : \sup_{i,j} |x_{ij}|/Q_{\varepsilon,j} > 1\}; \tag{22}$$

here $Q_{\varepsilon,j} = \varepsilon \sqrt{2(j + \log j) \log 2 + \log \log \varepsilon^{-1}}$.

On the other hand, for the regions of Gaussian asymptotics asymptotically minimax test procedures (for $h \leq p$, $t \geq q$) or minimax consistent tests are of the form $\psi_{\varepsilon,\alpha} = 1_{\mathcal{X}_\varepsilon \cup \{t_\varepsilon > T_\alpha\}}$. These are based on statistics

$$t_\varepsilon(x) = u_\varepsilon^{-1} \sum_{j=1}^{\infty} h_{\varepsilon,j} t_j(x_j, z_{\varepsilon,j}), \qquad t_j(x_j, z) = \sum_{i=1}^{2^j} \xi(x_{ij}/\varepsilon, z), \tag{23}$$

where

$$\xi(t, z) = \frac{1}{2}\left(\frac{dP_z}{dP_0}(t) + \frac{dP_{-z}}{dP_0}(t)\right) - 1 = e^{-z^2/2} \cosh(tz) - 1.$$

Here u_ε is determined by 11 and $h_{\varepsilon,j} \in (0,1]$, $z_{\varepsilon,j} \geq 0$ are families of sequences which correspond to asymptotically least favorable priors. For $p \leq 2$, $q \geq p$ we have $h_{\varepsilon,j} = 1$ and we can use statistics of χ^2-type

$$t_\varepsilon(x) = u_\varepsilon^{-1} \sum_{j=1}^{\infty} z^2_{\varepsilon,j} \chi^2(x_j), \qquad \chi^2(x_j) = \frac{1}{2} \sum_{i=1}^{2^j} (\varepsilon^{-2} x^2_{ij} - 1),$$

$$u^2_\varepsilon = \sum_{j=1}^{\infty} 2^{j-1} z^4_{\varepsilon,j}. \tag{24}$$

Applying wavelet transforms we obtain minimax consistent test procedures for alternatives 3, 5 under the Gaussian functional model. Since the rate asymptotics do not depend on parameters h, t (excepting boundary types of the Gaussian asymptotics), well known relationships between Sobolev and Besov norms lead to minimax consistent test procedures for alternatives which are determined by Sobolev norms for the regions of the main types of the Gaussian asymptotics.

5 Adaptive Setting

As it was noted above, the structure of asymptotical minimax or minimax consistent tests for the regions of Gaussian asymptotics is determined by families of sequences $\bar{h}_\varepsilon = \{h_{\varepsilon,j}\}$, $\bar{z}_\varepsilon = \{z_{\varepsilon,j}\}$. These families depend essentially on parameters $\tau = (\kappa, h, t)$, $\kappa = (\sigma, \eta, p, q)$ which determine norms 6 in the constraints 4. The dependence is not simple. Usually exact knowledge on these parameters is not available for the statistician.

The situation is analogous to that in estimation theory where the structure of the estimators which provide the minimax rates depends essentially on parameters which determine losses and constraints.

For these reasons we would like to construct tests which provide good minimax properties for wide enough regions of unknown parameters. First this problem has been studied by [20], [21]. Formal setting corresponds to alternatives of the form (we assume a quantity R be fixed)

$$\Theta_\varepsilon(\Gamma) = \bigcup_{\tau \in \Gamma} \Theta_\varepsilon(\tau, \rho_\varepsilon(\tau), R);$$

here the sets $\Theta_\varepsilon(\tau, \rho, R)$ are defined by 4 with the norms defined by 6; $\Gamma = K \times \Delta^2$, K and $\Delta^2 \subset R_+^2$ are compacts, $K \subset \Xi_{G_1} \cup \Xi_{G_2}$; the sets Ξ_{G_1}, Ξ_{G_2} are regions of parameters κ which correspond to the main types of Gaussian asymptotics and $\rho_\varepsilon(\tau)$ is a family of radii of removing balls. It means we would like to obtain good quality of testing *for all* $\tau \in \Gamma$.

Clearly, the functions $\rho_\varepsilon(\tau)$ should satisfy $\rho_\varepsilon(\tau)/\rho_\varepsilon^*(\kappa) \to \infty$ uniformly on Γ. It is the same that

$$u_\varepsilon(\Gamma) = \inf_{\tau \in \Gamma} u_\varepsilon(\tau, \rho_\varepsilon(\tau), R) \to \infty.$$

However it is not sufficient. Define *adaptive rate functions* $\rho_{\varepsilon,ad}^* = \rho_{\varepsilon,ad}^*(\kappa, R)$ by the relation

$$u_\varepsilon(\tau, \rho_{\varepsilon,ad}^*, R) = \sqrt{2 \log \log(R/\varepsilon)} + O(1);$$

where the quantities $u_\varepsilon(\tau, \rho_\varepsilon, R)$ are determined by 14 and, up to factor $d \sim 1$ these do not depend on h, t. Then we have the conditions for $\gamma_\varepsilon(\Theta_\varepsilon(\Gamma)) \to 1$ and $\gamma_\varepsilon(\Theta_\varepsilon(\Gamma)) \to 0$ analogous to 18, 19. These lead to losses in adaptive rates with respect to non-adaptive rates $\rho_\varepsilon = \rho_\varepsilon^*(\kappa)$ defined by 15 in the factor $(\log \log(R/\varepsilon))^{1/A_l(\kappa)}$, $\kappa \in \Xi_{G_l}$, $l = 1, 2$. One can interpret the losses as "price to pay for adaptation".

It is important that one can use "Bonferroni methods" to construct adaptive consistent tests families: it suffices to combine the thresholding 22 and collections of tests based on χ^2-statistics $t_{\varepsilon,0}$ of the form 24 and on the statistics $t_{\varepsilon,k}$ of the form 23 for specially selected sequences $\{z_{j,k}\}$, $k = 0, 1, ..., K_{e,j}$, $J_{0,\varepsilon} \leq j \leq J_\varepsilon \to \infty$, $J_{0,\varepsilon} = o(J_\varepsilon)$ and increased thresholds. Namely, we take $t_{j,0} = (2^{j+2} \log j)^{-1/2} \chi^2(x_j)$,

$$t_{j,k} = \left((2^{j+2} \sinh^2(z_{j,k}^2/2) \log j\right)^{-1/2} t_j(x_j, z_{j,k})$$

and we consider tests of the form

$$\psi_\varepsilon^{ad} = \mathbf{1}_{\mathcal{X}_\varepsilon}, \qquad \mathcal{X}_\varepsilon = \mathcal{X}_\varepsilon^{thr} \cup \{x : \max_{k,j} t_{j,k}(x) > 1\}.$$

For the cases $p \leq 2$, $q \geq p$ it suffices to consider combinations of collections of tests based on normalized statistics $t_{\varepsilon,0}$ of χ^2-type 24 only.

The wavelet transform provides the translation of results onto Besov and Sobolev norms in the functional space under the Gaussian functional model as well.

References

1. Burnashev, M.V. (1979). On the minimax detection of an inaccurately known signal in a Gaussian noise background. Theor. Probab. Appl. **24**, 107–119
2. Donoho, D.L. (1993). Asymptotic minimax risk for sup-norm loss: solution via optimal recovery. Probab. Theory Rel. **99**, 145–170
3. Donoho, D.L., Johnstone, I.M. (1992). Minimax estimation via wavelet shrinkage. Technical Report 402, Dep. of Statistics, Stanford University.
4. Donoho, D.L., Johnstone, I.M., Kerkyacharian, G., Picard, D. (1995). Wavelet shrinkage: asymptopia? (with discussion). J. Roy. Stat. Soc. B **57**, 301–369
5. Härdle, W., Kerkyacharian, G., Picard, D., Tsybakov, A. (1998). *Wavelets, Approximation, and Statistical Applications*. LNS,**129** Springer, New York
6. Ibragimov, I.A., Khasminskii, R.Z. (1977). One problem of statistical estimation in a white Gaussian noise. Sov. Math. Dokl. **236**, 333–337
7. Ibragimov, I.A., Khasminskii, R.Z. (1980). Asymptotic properties of some nonparametric estimates in a Gaussian white noise. In: *Proc. 3rd Summer School on Probab. Theory and Math. Stat. Varna 1978*, Sofia. 31–64
8. Ibragimov, I.A., Khasminskii, R.Z. (1980). On estimation of a probability density. Zapiski Nauchn. Seminar. LOMI. **98**, 66–85
9. Ibragimov, I.A., Khasminskii, R.Z. (1981). *Statistical Estimation: Asymptotic Theory*. Springer, Berlin-New York
10. Ingster, Yu.I. (1993). Asymptotically minimax hypothesis testing for nonparametric alternatives. I, II, III. Math. Method. Stat. **2**, 85–114, 171–189, 249–268
11. Ingster, Yu.I. (1998). Adaptation in minimax non-parametric hypothesis testing. WIAS, Preprint No. 419. Berlin.
12. Ingster, Yu.I., Suslina, I.A. (2000). Minimax nonparametric hypothesis testing for ellipsoids and Besov bodies. ESAIM: Probab. Stat. **4**, 53–135

13. Juditsky, A. (1997). Wavelet estimators: adapting to unknown smoothness. Math. Method. Stat. **6**, 1–25
14. Korostelev, A.P. (1993). Asymptotically minimax regression estimator in uniform norm up to exact constant. Theor. Probab. Appl. **38**, 737–743
15. Lepski, O.V., Mammen, E., Spokoiny, V.G. (1997). Optimal spatial adaptation to inhomogeneous smoothness: an approach based on kernel estimates with variable bandwidth selectors. Ann. Stat. **25**, 929–947
16. Lepski, O.V., Spokoiny, V.G. (1997). Optimal pointwize adaptive methods in nonparametric estimation. Ann. Stat. **25**, 2512–2546
17. Lepski, O.V., Spokoiny, V.G. (1999). Minimax nonparametric hypothesis testing: the case of an inhomogeneous alternative. Bernoulli **5**, 333–358
18. Lepski, O.V., Tsybakov, A.B. (2000). Asymptotically exact nonparametric hypothesis testing in sup-norm and at a fixed point. Probab. Theory Rel. **117**, 17–48
19. Pinsker, M.S. (1980). Optimal filtration of square-integrable signals in Gaussian noise. Probl. Inf. Transm. **16**, 120–133
20. Spokoiny, V.G. (1996). Adaptive hypothesis testing using wavelets. Ann. Stat. **24**, 2477–2498
21. Spokoiny, V.G. (1998). Adaptive and spatially adaptive testing of nonparametric hypothesis. Math. Method. Stat. **7**, 245–273
22. Triebel H. (1992). *Theory of Functional Space* 2. Birkhäuser, Basel.

Testing Randomness on the Basis of the Number of Different Patterns

Andrew L. Rukhin

Department of Mathematics and Statistics, University of Maryland Baltimore County, Baltimore, USA

Abstract. The problem of randomness testing gained importance because of the need to assess the quality of different random number generators. The wide use of public key cryptography necessitated testing for randomness binary strings produced by such generators. The evaluation of random nature of various generators outputs became vital for communications industry where digital signatures and key management are crucial for information processing and for computer security.

The tests discussed here are based on the observed numbers of patterns which appear with a given frequency. Our results are based on some properties of the so-called pattern correlation matrices which are useful in statistical analysis of random sequences.

1 Introduction and Summary

Consider a random text formed by realizations of letters chosen from a finite alphabet. For a given set of patterns it is of interest to determine the probability of the prescribed number of occurrences of the patterns in the text. This problem appears in different areas of information theory like source coding, code synchronization, randomness testing, etc. It also is important in molecular biology in DNA analysis and for gene recognition.

The wide use of public key cryptography makes it necessary to test for randomness strings produced by such generators. Also common secure encryption algorithms are based on a generator of (pseudo) random numbers. The testing of such generators for randomness is quite important for communications industry.

A number of classic tests of randomness are reviewed in Knuth ([4]). However, some of these tests pass patently nonrandom sequences a(see discussion in Marsaglia, [6]). The most popular collection of tests for randomness, the Diehard Battery, demands fairly long strings (up to 2^{24} bits). A commercial product, called CRYPT-X, (Gustafson et al. [3]) includes some of tests for randomness. A more recent battery of randomness tests is described in Rukhin ([10]).

Since many conventional pseudo random numbers generators because of their deterministic recursive algorithms exhibit patterned outputs, it is natural to employ statistical tests based on the occurrences of patterns (patterns) of a given length. The test discussed here utilizes the observed numbers of patterns which appear a given number of times (i.e. which are missing, appear exactly once, exactly twice, etc.) The number of missing two-letter patterns is used in the "OPSO Theory" which is a part of the Diehard Battery (Marsaglia ([7])). Also a similar test has been investigated by Tikhomirova and Chistyakov ([12]). The usefulness of similar procedures in quality control was demonstrated by Shmueli and Cohen ([11]).

In this paper in Section 2 we derive the necessary results for the generating functions of probabilities for patterns from a given collection to occur in random text a given number of times. These results extend the previous formulas obtained by Guibas and Odlyzko ([2]), and by Fudos, Pitoura and Szpankowski ([1]). Section 3 deals with the expected value of the number of patterns occurring a given number of times, and the covariance structure of the corresponding random variables. In Section 4 the optimal linear test based on these statistics is given.

2   Correlation Polynomials and Generating Functions

Denote by $\epsilon_1, \ldots, \epsilon_n$ a sequence of i.i.d. random variables each taking values in the finite set $\{1, \ldots, q\}$ such that $P(\epsilon_i = \ell) = p_\ell, \ell = 1, \ldots, q$. We will be interested in occurrences of patterns of the form $i = (i_1 \ldots i_m)$ whose probability is $P(i) = p_{i_1} \cdots p_{i_m}$. The situation when $p_\ell \equiv q^{-1}$ corresponds to the randomness hypothesis. We will need the probability $\pi_i^r(n)$ that a fixed pattern i appears in the string of length n exactly r times.

The following *correlation polynomial* of two patterns is needed to obtain the distribution of the numbers of patterns occurring with a given frequency. Let $i = (i_1 \cdots i_m)$ and $j = (j_1 \ldots j_m)$ be two patterns of length m. Put

$$C_{ij}(z) = \sum_{k=1}^{m} \delta_{(i_{m-k+1} \cdots i_m),(j_1 \ldots j_k)} p_{j_{k+1}} \cdots p_{j_m} z^{k-1}. \tag{1}$$

We denote by $C(z)$ the *correlation matrix*,

$$C(z) = \begin{pmatrix} C_{ii}(z) & C_{ij}(z) \\ C_{ji}(z) & C_{jj}(z) \end{pmatrix}.$$

For *aperiodic* patterns i of length m, $C_{ii}(z) = z^{m-1}$.

According to Theorem 3.3 in Guibas and Odlyzko (1981) the probability generating function $F_i^0(z) = \sum_n \pi_i^0(n) z^{-n}$ of probabilities $\pi_i^0(n)$ $= P(\text{pattern } i \text{ is missing})$, has the form

$$F_i^0(z) = \frac{z C_{ii}(z)}{B(z)}, \tag{2}$$

where

$$B(z) = B_i(z) = (z-1)C_{ii}(z) + P(i) = \prod_{j=1}^{m}(z - z_j).$$

For $r \geq 1$ a formula for the generating function $F_i^r(z)$ for probabilities $\pi_i^r(n)$ has been reported by Fudos, Pitoura and Szpankowski (1995). It has the form

$$F_i^r(z) = \frac{z^m P(i)[(z-1)(C_{ii}(z) - z^{m-1}) + P(i)]^{r-1}}{[B(z)]^{r+1}}. \tag{3}$$

The probabilities of the form, $\pi_{ij}^{00}(n) = P \,(\text{patterns } i \text{ and } j \text{ are missing})$, also can be determined from the generating function the form of which follows from the simultaneous equations of Theorem 3.3, p 195 in Guibas and Odlyzko ([2]). These equations are applicable for any finite set of different patterns $\Omega = \{A, B, \ldots, T\}$ (each of positive probability). According to this theorem the generating function for probabilities of two specific patterns i and j to be missing in the string of length n depends on the correlation matrix $\mathcal{C}(z)$ in the following way

$$F_{ij}^{00}(z) = \sum_n \frac{P(\text{patterns } i, j \text{ are missing in } n\text{-string})}{z^n}$$

$$= \frac{z|\mathcal{C}(z)|}{(z-1)|\mathcal{C}(z)| + P(j)C_{ii}(z) + P(i)C_{jj}(z) - P(i)C_{ij}(z) - P(j)C_{ji}(z)}.$$

Here $|\mathcal{C}(z)| = C_{ii}(z)C_{jj}(z) - C_{ij}(z)C_{ji}(z)$ denotes the determinant of $\mathcal{C}(z)$.

We give now an extension of this formula to the generating function for the probabilities $\pi_{ij}^{rt}(n)$, that a given pattern i occurs in the string of length n exactly r times and a pattern j occurs t times.

These probabilities refer to a random string ϵ of length n and to a set $\Omega = \{A, B, \ldots, T\}$ of patterns which have the same length m. Put

$$P_{GH}(n) \tag{4}$$

$= P \,(G\epsilon \text{ ends with } H \text{ and } G\epsilon \text{ does not contain any other patterns from } \Omega \,).$

Observe that when $G = \emptyset$,

$P_{\emptyset H}(n) = P \,(H \text{ is the only pattern from } \Omega, \text{ appearing at the end of } \epsilon \,).$

When $H = \emptyset$, $P_{G\emptyset}(n) = P \,(\text{the only pattern from } \Omega \text{ in } \epsilon G \text{ is } G)$. Observe that $P_{G\emptyset}(0) = 1$, but $P_{GH}(0) = 0$ for $G, H \in \Omega$. Also if $P_{GH}(n) > 0$, then for any $\ell = 1, \ldots, m - n$, $(G_{m-\ell+1} \cdots G_m) = (H_{m-\ell-n+1} \ldots H_{m-n})$, so that $P_{GH}(n) = \delta_{(G_{m-\ell+1} \cdots G_m),(H_{m-\ell-n+1} \ldots H_{m-n})} P_{GH}(n)$.

Theorem 1. *For any $A \in \Omega$ the generating functions F_{GH} of probabilities $P_{GH}(n)$ in (4) satisfy the following simultaneous equations*

$$(z-1)F_{A\emptyset}(z) + zF_{AA}(z) + \cdots + zF_{AT}(z) = z$$

$$P(A)F_{A\emptyset}(z) - zC_{AA}(z)F_{AA}(z) - \cdots - zC_{TA}(z)F_{AT}(z) = z^m - zC_{AA}(z)$$

$$\cdots \cdots \cdots \cdots \cdots \cdots \cdots$$

$$P(H)F_{A\emptyset}(z) - zC_{AH}(z)F_{AA}(z) - \cdots - zC_{TH}(z)F_{AT}(z) = -zC_{AH}(z)$$

$$\cdots \cdots \cdots \cdots \cdots \cdots \cdots$$

$$P(T)F_{A\emptyset}(z) - zC_{AT}(z)F_{AA}(z) - \cdots - zC_{TT}(z)F_{AT}(z) = -zC_{AT}(z). \quad (5)$$

As the main Corollary of Theorem 1 we give the form of the generating functions for a two-element set Ω.

Theorem 2. *Let $\Omega = \{i, j\}$ with positive $P(i), P(j)$. Then*

$$F_{i\emptyset}(z) = \frac{z^m(C_{jj}(z) - C_{ij}(z))}{B_{ij}(z)} = \frac{P_{i\emptyset}(z)}{B_{ij}(z)},$$

$$F_{\emptyset i}(z) = \frac{P(i)C_{jj}(z) - P(j)C_{ji}(z)}{B_{ij}(z)} = \frac{P_{\emptyset i}(z)}{B_{ij}(z)},$$

$$F_{ij}(z) = \frac{z^{m-1}((z-1)C_{ij}(z) + P(j))}{B_{ij}(z)} = \frac{P_{ij}(z)}{B_{ij}(z)},$$

$$F_{ii}(z) = 1 - \frac{z^{m-1}((z-1)C_{jj}(z) + P(j))}{B_{ij}(z)} = \frac{P_{ii}(z)}{B_{ij}(z)},$$

with similar formulas for $F_{j\emptyset}, F_{\emptyset j}, F_{ji}$ and F_{jj}.

Theorem 2 leads to an explicit form of the generating function $F_{ij}^{rt}(z)$ for the probabilities $\pi_{ij}^{rt}(n)$ of patterns i and j to occur in the string of length n, r and t times respectively. In particular, it gives the desired formula for the case $r = t = r$.

Theorem 3. *For $r \wedge t \geq 1$ one has*

$$F_{ij}^{rt}(z) = \frac{1}{[B_{ij}(z)]^{r+t+1}} \sum_{k=0\vee(r-t-1)}^{r-2} \binom{r-1}{k}\binom{t-1}{r-k-2}$$

$$\times [P_{ii}(z)]^k [P_{jj}(z)]^{t-r+k+1} [P_{ij}(z)P_{ji}(z)]^{r-k-1} P_{\emptyset i}(z)P_{i\emptyset}(z)$$

$$+ \sum_{k=0\vee(r-t+1)}^{r-1} \binom{r-1}{k}\binom{t-1}{r-k} [P_{ii}(z)]^k [P_{jj}(z)]^{t-r+k-1}$$

$$\times [P_{ij}(z)P_{ji}(z)]^{r-k} P_{\emptyset j}(z)P_{j\emptyset}(z)$$

$$+ \sum_{k=0\vee(r-t)}^{r-1} \binom{r-1}{k} \binom{t-1}{r-k-1} [P_{ii}(z)]^k [P_{jj}(z)]^{t-r+k} [P_{ij}(z)P_{ji}(z)]^{r-k-1}$$

$$\times [P_{\emptyset i}(z)P_{ij}(z)P_{j\emptyset}(z) + P_{\emptyset j}(z)P_{ji}(z)P_{i\emptyset}(z)]. \tag{6}$$

Also

$$F_{ij}^{r0}(z) = \frac{1}{[B_{ij}(z)]^{r+1}} [P_{ii}(z)]^{r-1} P_{\emptyset i}(z)P_{i\emptyset}(z).$$

When $r = t = r$

$$F_{ij}^{rr}(z) = \frac{1}{[B_{ij}(z)]^{2r+1}}$$

$$\sum_{k=0}^{r-1} \binom{r-1}{k}^2 [P_{ii}(z)P_{jj}(z)]^k [P_{ij}(z)P_{ij}(z)]^{r-k-1}$$

$$\times [P_{\emptyset i}(z)P_{ij}(z)P_{j\emptyset}(z) + P_{\emptyset j}(z)P_{ji}(z)P_{i\emptyset}(z)]$$

$$+ \sum_{k=0}^{r-2} \binom{r-1}{k} \binom{r-1}{k+1} [P_{ii}(z)P_{jj}(z)]^k [P_{ij}(z)P_{ij}(z)]^{r-k-1}$$

$$\times [P_{\emptyset i}(z)P_{jj}(z)P_{i\emptyset}(z) + P_{\emptyset j}(z)P_{ii}(z)P_{j\emptyset}(z)]. \tag{7}$$

3   Asymptotic Formulas for the Expected Number of Patterns Appearing a Given Number of Times and for Their Covariances

The formulas for the generating functions discussed in the previous Section lead to the asymptotic behavior of the first two moments of the number of patterns appearing a given number of times. In this asymptotic study it will be assumed that as $n \to \infty$, $n/q^m \to \alpha$ with a fixed positive α. To study asymptotic efficiency of tests for randomness, we consider the case when the distribution of the alphabet letters is close to the uniform in the sense that $p_\ell = q^{-1} + q^{-3/2}\eta_\ell$ with $\sum_{k=1}^q \eta_k = 0$. It will be assumed that as $n \to \infty$, $q^{-1} \sum_k \eta_k^2 \to \mathbf{D} > 0$.

Let $s, s \geq 1$, denote the *period* of i, i.e. the smallest positive integer for which $(i_{s+1} \cdots i_m) = (i_1 \dots i_{m-s})$. Let $P(i_s) = p_{i_{m-s+1}} \cdots p_{i_m}$. If i is an aperiodic word, i.e. $C_{ii}(z) = z^{m-1}$, then there is no such positive integer, and we put $s = \infty$ with $P(i_\infty) = 0$.

Under our convention about the probabilities of the letters, one has

$$P(i_s) = \frac{1}{q^s} + \frac{\eta_{i_{m-s+1}} + \cdots + \eta_{i_m}}{q^{s+1/2}} + \frac{\sum_{m-s+1 \leq k < j \leq m} \eta_{i_k}\eta_{i_j}}{q^{s+1}} + O\left(\frac{1}{q^{s+3/2}}\right)$$

and

$$P(i) = \frac{1}{q^m} + \frac{\eta_{i_1} + \cdots + \eta_{i_m}}{q^{m+1/2}} + \frac{\sum_{1 \leq k < j \leq m} \eta_{i_k}\eta_{i_j}}{q^{m+1}} + O\left(\frac{1}{q^{m+3/2}}\right).$$

The polynomial $B(z)$ in (2) has the form

$$B(z) = z^{m-1}(z-1) + P(i_s)z^{m-s-1}(z-1)$$

$$+ \sum_{k=1}^{m-s-1} \delta_{(i_{m-k+1}\cdots i_m),(i_1\ldots i_k)} p_{i_{k+1}} \cdots p_{i_m} z^{k-1}(z-1) + P(i) = \prod_{j=1}^{m}(z - z_j).$$

It is not difficult to show that the largest root z_1 of the equation $B(z) = 0$ is real, $z_1 < 1$, $z_1 \to 1$ as $q \to \infty$, and all other roots tend to zero, $\max_{j\geq 2}|z_j| \leq q^{-1}$.

By using this facts one obtains for $r \geq 1$

$$\pi_i^r(n) = \frac{\alpha^r e^{-\alpha}}{r!}\left[1 + \frac{(r-\alpha)\sum_k \eta_{i_k}}{q^{1/2}}\right.$$

$$+ \frac{(r(r-1) - 2\alpha r + \alpha^2)\left(\sum \eta_{i_k}\right)^2 + 2(r-\alpha)\sum_{1\leq k<j\leq m}\eta_{i_k}\eta_{i_j}}{2q}$$

$$\left. + \frac{r(r-1) - 2\alpha r + \alpha^2}{\alpha q^s} + O\left(\frac{1}{q^{3/2}}\right)\right]. \tag{8}$$

When $\eta_i \equiv 0$,

$$\pi_i^r(n) = \frac{\alpha^r e^{-\alpha}}{r!}\left[1 + \frac{r(r-1) - 2\alpha r + \alpha^2}{\alpha q^s} + o\left(\frac{1}{q^s}\right)\right],$$

and for aperiodic patterns i

$$\pi_i^r(n) = \frac{\alpha^r e^{-\alpha}}{r!}$$

$$\times \left[1 - \frac{1}{q^m}\left(\left(m - \frac{1}{2}\right)\alpha - m(2r+1) + r + 1 + \frac{r(4mr + 2m - 3r - 3)}{2\alpha}\right)\right.$$

$$\left. + O\left(\frac{1}{q^{2m}}\right)\right].$$

Also

$$\pi_i^0(n) = e^{-\alpha}$$

$$\times \left[1 + \frac{\alpha \sum_k \eta_{i_k}}{q^{1/2}} + \frac{\alpha^2\left(\sum \eta_{i_k}\right)^2 - 2\alpha \sum_{1\leq k<j\leq m}\eta_{i_k}\eta_{i_j}}{2q}\right.$$

$$\left. + \frac{\alpha+1}{q^s} + O\left(\frac{1}{q^{3/2}}\right)\right]. \tag{9}$$

As in the previous case, for the uniform distribution on the alphabet and periodic patterns i with period s

$$\pi_i^0(n) = e^{-\alpha}\left[1 + \frac{\alpha}{q^s} + O\left(\frac{1}{q^{2s\wedge m}}\right)\right]$$

and for aperiodic patterns

$$\pi_i^0(n) = e^{-\alpha}\left[1 - \frac{(2m-1)\alpha}{2q^m} + \frac{m-1}{q^m} + O\left(\frac{1}{q^{2m}}\right)\right].$$

The form of the probabilities (8) and (9) leads to the formula for the expected value of the number of m-patterns, which occur exactly r times in a string of length n, $X^r = X_n^r$.

As $\sum_{k<j}\eta_{i_k}\eta_{i_j} = 0$, $\sum_i(\sum_k\eta_{i_k})^2 = \sum_i\sum_k\eta_{i_k}^2 = mq^{m-1}\sum_{\ell=1}^q\eta_k^2 = mq^m\mathbf{D}$. Therefore, with $\pi_i(n)$ determined from (9), one gets for any $r = 0, 1, \ldots$

$$\mathbf{E}X^r = \sum_i \pi_i^r(n)$$

$$= \frac{\alpha^r e^{-\alpha}}{r!}q^m\left[1 + \frac{m\mathbf{D}}{2q}\left[\alpha^2 - 2\alpha r + r(r-1)\right] + O\left(\frac{1}{q^{3/2}}\right)\right]. \quad (10)$$

Actually, when $\eta_i \equiv 0$, one can get a much more accurate asymptotic formula. Let $Q_s = Q_s(m), s = 1, \ldots, m-1, \infty$ denote the total number of patterns whose correlation polynomial has the form (1) (with $s = \infty$ corresponding to aperiodic patterns). Then

$$\sum Q_s = q^m,$$

and one can prove that as $q \to \infty$ for $s = 1, \ldots, m-1$

$$\frac{Q_s}{q^s} \to 1, \qquad \frac{Q_\infty}{q^m} \to 1.$$

As the asymptotic behavior of π_i^r is determined by the period of the pattern i, these formulas imply that for any fixed $r = 0, 1, \ldots$

$$\mathbf{E}X^r = \sum_s Q_s\pi_i^r(n)$$

$$= \frac{\alpha^r e^{-\alpha}}{r!}\left[q^m - \frac{\alpha}{2} + m + r - 1 - \frac{r(2mr + 4m - r - 5)}{2\alpha}\right] + O\left(\frac{1}{q}\right). \quad (11)$$

The derivation of the asymptotic formula for the covariance is more cumbersome. To obtain it, note that $X^r = \sum_j x_j^r$ where x_j^r is 0 or 1 according to occurrence of the pattern j in the string of length n exactly r times. Thus, $\mathbf{E}x_i^r x_j^t = \pi_{ij}^{rt}(n) = P(i$ appears r times, j appears t times$)$, so that

$$\mathbf{Var}(X^r) = \sum_i \mathbf{Var}(x_i) + \sum_{i\neq j} \mathbf{Cov}(x_i, x_j)$$

$$= \sum_i \pi_i^r(n)[1 - \pi_i^r(n)] + \sum_{i\neq j}[\pi_{ij}^{rr}(n) - \pi_i^r(n)\pi_j^r(n)] \quad (12)$$

and for $r \neq t$

$$\text{Cov}(X^r, X^t) = \sum_{i \neq j}[\pi_{ij}^{rt}(n) - \pi_i^r(n)\pi_j^t(n)]. \qquad (13)$$

The probabilities $\pi_i^r(n)$ have been determined in (9) and (8). The remaining probabilities $\pi_{ij}^{rt}(n)$ can be found from the generating functions in Theorem 3. They are determined by the periods s and u of the autocorrelation polynomials $C_{ii}(z)$ and $C_{jj}(z)$, and by the degrees $m - 1 - v$ and $m - 1 - w$ of the polynomials $C_{ij}(z)$ and $C_{ji}(z)$ respectively. (Then for $i \neq j$, $C_{ij}(z) = z^{m-1-v}P(j_v) + \cdots + \delta_{i_m j_1}P(j_{m-1})$ and $C_{ji}(z) = z^{m-1-w}P(i_w) + \cdots + \delta_{i_1 j_m}P(i_{m-1})$ with $1 \leq v, w \leq m - 1$.) One can show that

$$\pi_{ij}^{rt}(n) - \pi_i^r(n)\pi_j^t(n)$$

$$= \frac{e^{-2\alpha}\alpha^{r+t}}{r!t!}\left[\left(\alpha + r - 1 + \frac{r(r-1)}{\alpha}\right)P(i_s) + \left(\alpha + t - 1 + \frac{t(t-1)}{\alpha}\right)P(j_u)\right.$$

$$\left. + (\alpha - r - t)(P(j_v) + P(i_w))\right]\left[1 + O\left(\frac{1}{q}\right)\right]. \qquad (14)$$

The main term of the asymptotic expansion of $\text{Var}(X^r)$ and $\text{Cov}(X^r, X^t)$ is determined by the sums of terms proportional to $P(i_s), P(j_u), P(j_v)$ and $P(i_w)$. Denote by Q_{su}^{vw} the number of pairs of different patterns i, j for which the autocorrelation polynomials have periods s and u, and the correlation polynomials $C_{ij}(z)$ and $C_{ji}(z)$ have the degrees $m - 1 - v$ and $m - 1 - w$ respectively. Clearly,

$$Q_{su}^{vw} \sim \begin{array}{ll} q^{s+u} & s \wedge u < \infty \\ q^{m+u} & s = \infty, u < \infty \\ q^{m+v} & s = u = \infty, v < \infty \\ q^{2m} & s = u = v = w = \infty \end{array}$$

It follows that the main contribution to the sum in (12) is due to the pairs of aperiodic patterns such that at least one of the polynomials $C_{ij}(z)$ or $C_{ji}(z)$ vanishes. Thus one can put $\eta_i \equiv 0$.

One gets for $r \neq t$

$$\text{Cov}(X^r, X^t) = \sum Q_{su}^{vw}[\pi_{ij}^{rt}(n) - \pi_i^r(n)\pi_j^t(n)] - \sum Q_s\pi_i^r(n)\pi_i^t(n)$$

$$= -q^m\frac{e^{-2\alpha}\alpha^{r+t}}{r!t!}\left[(2m-1)\alpha - (r+t)(2m-1) + 1\right.$$

$$\left. + \frac{rt}{\alpha}(4m-3) - 2(m-1)\frac{(\alpha - r)(\alpha - t)}{\alpha}\right] + O\left(q^{m-1}\right)$$

$$= -q^m\frac{e^{-2\alpha}\alpha^{r+t}}{r!t!}\left[\alpha - r - t + 1 + \frac{rt}{\alpha}\right] + O\left(q^{m-1}\right). \qquad (15)$$

Similarly

$$\mathbf{Var}(X^r) = \sum_s Q_s \pi_i^r(n)[1 - \pi_i^r(n)] + \sum Q_{su}^{vw}[pi_{ij}^{rr}(n) - \pi_i^r(n)\pi_j^r(n)]$$

$$= q^m \frac{e^{-\alpha}\alpha^r}{r!}\left[1 - \frac{e^{-\alpha}\alpha^r}{r!}\left(\alpha - 2r + 1 + \frac{r^2}{\alpha}\right)\right] + O\left(q^{m-1}\right). \quad (16)$$

Theorem 4. *The probability* $\pi_i^r(n)$ *admits the asymptotic representations (9) and (8) when* $n \to \infty$, $n/q^m \to \alpha$ *with a fixed positive* α. *The expected value of the number of patterns appearing exactly* r *times in a random text has the form (10) with a more accurate formula (11) in the case* $\eta_i \equiv 0$. *The covariance between the number of patterns appearing exactly* r *and* t *times has the form (15), and the variance of the number of patterns appearing exactly* r *times has the form (16).*

Kolchin, Sevastyanov and Chistyakov ([5]) in Theorem 6 of Chapter III give the formulas for the first two moments of the joint distribution of the number of patterns appearing a prescribed number of times when the occurrences of these patterns are independent, i.e. when the patterns appearances in the non-overlapping m-blocks are counted. A rather surprising fact is that the asymptotic behavior of the expected value and of the covariance matrix is the same for overlapping and non-overlapping occurrences.

4   Asymptotic Normality and the Optimal Linear Test

The theoretical justification for approximate normality of the distribution of X^r when $n \to \infty$, $n/q^m \sim \alpha$ follows from a limit theorem by Mikhailov ([9]). According to Theorem 4, $\mathbf{Var}(X^r) \to \infty$, so that the crucial condition in Mikhailov's theorem is satisfied.

For a fixed positive integer R, denote by Σ the covariance matrix of the limiting distribution of the random variables $X^0, X^1, \ldots, X^R$. According to the next theorem this distribution is normal.

Theorem 5. *The random number of m-patterns,* $X^r = X_n^r$, *which appear exactly* r *times in a string of length* n *when* $n \to \infty$, $n/q^m \to \alpha$ *with a fixed positive* α, *is asymptotically normal. If the probability of the i-th letter* $(i = 1, \ldots, q)$ *is of the form* $q^{-1} + \eta_i q^{-3/2}$, $\sum_{k=1}^q \eta_k = 0$, *then the asymptotic mean is given by (10) with* $\mathbf{D} = q^{-1}\sum_k \eta_k^2$, *and by (11) for the uniform distribution. In both cases the variance is determined from (16). The asymptotic joint distribution of the random variables* $X^0, X^1, \ldots, X^R$ *is normal with the covariance matrix* Σ.

With $p_r(\alpha) = \frac{\alpha^r e^{-\alpha}}{r!}$ denoting the Poisson probabilites, the elements of matrix Σ have the form

$$\sigma_{rr} = p_r(\alpha)\left[1 - p_r(\alpha)\left(\alpha - 2r + 1 + \frac{r^2}{\alpha}\right)\right],$$

and for $r \neq t$

$$\sigma_{rt} = -p_r(\alpha)p_t(\alpha) \left[\alpha - r - t + 1 + \frac{rt}{\alpha}\right]$$

$$= -p_r(\alpha)p_t(\alpha) \left[1 + \frac{(\alpha - r)(\alpha - t)}{\alpha}\right].$$

Thus with a diagonal matrix D formed by elements $p_r(\alpha)$, and

$$U^T = \begin{pmatrix} p_0(\alpha), \ldots, p_R(\alpha) \\ \sqrt{\alpha}p_0(\alpha), \ldots, p_R(\alpha)(\alpha - R)/\sqrt{\alpha} \end{pmatrix}$$

being of rank two, one obtains the following representation

$$\Sigma = D - UU^T. \tag{17}$$

We use Theorem 5 to derive the optimal test of the null hypothesis H_0 : $\eta_i \equiv 0$ within the class of linear statistics of the form

$$S = \sum_{r=0}^{R} w_r X^r$$

with some constants w_r.

According to Theorem 5 S is asymptotically normal both under the null hypothesis and the alternative $H_1 : \mathbf{D} > 0$. The Pitman efficiency of this statistic is determined by its efficacy, i.e. by the difference between the means under the null hypothesis and the alternative, divided by the standard deviation (which is common to the null hypothesis and the alternative),

$$\mathbf{eff}(S) = \frac{m| \sum_{r=0}^{R} w_r p_r(\alpha) \left[\alpha^2 - 2\alpha r + r(r-1)\right]|}{2 \left(\sum_{r,t} \sigma_{rt} w_r w_t\right)^{1/2}} = \frac{m|\mathbf{wb}^T|}{2\sqrt{\mathbf{w}^T \Sigma \mathbf{w}}}.$$

Here $(R + 1)$-dimensional vector $\mathbf{w}$ has coordinates $w_0, \ldots, w_R$ and $\mathbf{b}$ has coordinates $p_r(\alpha)(\alpha^2 - 2\alpha r + r(r-1)) = \alpha^2[p_r(\alpha) - 2p_{r-1}(\alpha) + p_{r-2}(\alpha)], r = 0, 1, \ldots, R$.

Maximization of this ratio leads to the solution

$$\mathbf{w} = \Sigma^{-1}\mathbf{b}$$

(or a scalar factor thereof.) Because of (17) the inverse of the matrix Σ has the following form

$$\Sigma^{-1} = D^{-1} + D^{-1}U \left[\begin{pmatrix} 1 & 0 \\ 0 & 1 \end{pmatrix} - U^T D^{-1} U\right]^{-1} U^T D^{-1},$$

With $\theta = \theta_R = p_R(\alpha) \left[\sum_{R+1} p_r(\alpha) \right]^{-1}$ and $D = 1 + (R - \alpha + 1)\theta - \alpha\theta^2$, this representation leads to the formula for the coordinates of $\mathbf{w}$

$$\mathbf{w}_r = \alpha^2 - 2\alpha r + r(r-1) + (\alpha - r)\theta \frac{[(\alpha - R)^2 + \alpha\theta(\alpha - R) + R]}{D}$$
$$+ \frac{(\alpha - R + \alpha\theta)\alpha\theta}{D}. \tag{18}$$

This formula can be used to show that for the value of $\alpha = \alpha^*(R)$, which maximizes the efficacy, the following limits exists

$$\lim_{R \to \infty} \frac{\alpha^*(R)}{R} = 1 \tag{19}$$

and for $\alpha = \alpha^*(R)$

$$\lim_{R \to \infty} \frac{\text{eff}(S)}{R} = \frac{m}{\sqrt{2}}. \tag{20}$$

Theorem 6. *The weights w_r of the optimal linear test statistic S are given by (18). The value of $\alpha = \alpha^*(R)$, which maximizes the efficacy, has the asymptotic expansion (19), and the limiting formula (20) holds.*

Following Section 3 of Chapter V of Kolchin, Sevastyanov and Chistyakov ([5]), one can show that the corresponding statistic is asymptotically optimal not only within the class of linear functions, but also in the class of all statistics of $X^0, \ldots, X^R$.

4.1   Example: Two-Letter Patterns

The number of missing pairs has been used by Tikhomirova and Chistyakov ([12]). This statistic is also employed in the so-called "OPSO Theory" introduced in Marsaglia ([6]) and used in the OPSO test of randomness in the Diehard Battery (Marsaglia [7]). In this test one takes non-overlapping substrings formed by zeros and ones of given length p to represent the letters of the new alphabet, so that there are $q = 2^p$ new letters. In OPSO test one counts the number of two-letter patterns (the original substrings of length $2p$) which never occurred. (In the Diehard test $p = 10, q = 2^{10}$.)

If the probabilities of alphabet letters have the form $\pi_i = q^{-1} + q^{-3/2}\eta_i$, then for aperiodic patterns $j = (A, B), A \neq B$,

$$F_j^0(z) = \frac{z^2}{p_A p_B + (z-1)z}.$$

With $s = \frac{1}{2} + \sqrt{\frac{1}{4} - p_A p_B}$, $t = \frac{1}{2} - \sqrt{\frac{1}{4} - p_A p_B}$,

$$
\begin{aligned}
\pi_{A,B} &= \frac{s^{n+1} - t^{n+1}}{s - t} \\
&= e^{-\alpha} - \frac{\alpha e^{-\alpha}}{\sqrt{q}}(\eta_A + \eta_B) - \frac{\alpha e^{-\alpha}}{q}\eta_A \eta_B + \frac{\alpha^2 e^{-\alpha}}{2q}(\eta_A + \eta_B)^2 \\
&\quad + \frac{\alpha^2 e^{-\alpha}}{q^{3/2}}\left[\eta_A \eta_B^2 + \eta_A^2 \eta_B - \frac{\alpha}{6}(\eta_A + \eta_B)^2\right] \\
&\quad + \frac{e^{-\alpha}}{q^2}\left[1 - \frac{3\alpha}{2} + \frac{\alpha^2}{2}\eta_A^2 \eta_B^2 - \frac{\alpha^3}{2}\eta_A \eta_B(\eta_A + \eta_B)^2 + \frac{\alpha^4}{24}(\eta_A + \eta_B)^4\right] \\
&\quad + O\left(\frac{1}{q^{5/2}}\right).
\end{aligned}
\tag{21}
$$

For periodic templates, $j = (A, A)$, a similar formula for the generating function shows that

$$
\pi_{A,A} = e^{-\alpha} - \frac{2\alpha e^{-\alpha}}{\sqrt{q}}\eta_A + \frac{\alpha e^{-\alpha}}{q}\left[1 + \alpha(2\alpha - 1)\eta_A^2\right] + O\left(\frac{1}{q^{3/2}}\right).
\tag{22}
$$

The approximate formulas (21) and (22) lead to very accurate answers for the expected value and the variance. The calculations according to the formulas above show that in Marsaglia's example when when $n = 2^{21}$, $q = 2^{10}$ (so that $\alpha = 2$),

$$\pi_{(A,A)} = 0.13559935200020, \quad \pi_{(A,B)} = 0.13533502510527.$$

The worst is approximation in (22) which gives the value

$$\pi_0(n) = 0.13559922276433.$$

For example, while the exact value of the mean is $EX = 141909.3299555$, the value determined from (10) is 141909.3299551.

References

1. Fudos, I., Pitoura, E., Szpankowski, W. (1996). On pattern occurrences in a random text. Inform. Process. Lett. **57**, 307–312
2. Guibas, L.J., Odlyzko, A.M. (1981). String overlaps, pattern matching and nontransitive games. J. Comb. Theory A, **30**, 183–208
3. Gustafson, H., Dawson, E., Nielsen, L., Caelli, W. (1994). A computer package for measuring the strength of encryption algorithms. IFIP Trans. A **13**, 687–697
4. Knuth, D.E. (1998). *The Art of Computer Programming*, Vol. 2, 3rd ed. Addison-Wesley Inc., Reading, MA
5. Kolchin, V.F., Sevast'yanov, B.A., Chistyakov, V.P. (1978). *Random Allocations*. Whinston Sons, Washington, DC

6. Marsaglia, G. (1985). A Current View of Random Number Generation. In *Computer Science and Statistics: Proceedings of the Sixteenth Symposium on the Interface*, 3–10 Elsevier Science Pub., New York

7. Marsaglia, G. (1996). *Diehard: A battery of tests for randomness.* http://stat.fsu.edu/ geo/diehard.html.

8. Menezes, A.J., van Oorschot, P.C., Vanstone, S.A. (1997). *Handbook of Applied Cryptography.* CRC Press, Boca Raton, FL

9. Mikhailov, V.G. (1991). Asymptotic normality of decomposable statistics from the frequencies of m-chains. Discrete Math. Appl. 1, 335–3471

10. Rukhin, A.L. (2000). Testing randomness: A suite of statistical procedures. Theor. Probab. Appl. **45**, 111–132

11. Shmueli, G., Cohen, A. (2000). Run-related probability functions applied to sampling inspection. Technometrics **46**, 2188–202

12. Tikhomirova, M.I., Chistyakov, V.P. (1997). On statistical tests based on missing s–patterns (in Russian). In: *Trydy po Diskretnoi Matematike*, TVP Publishers, Moscow

The π^* Index as a New Alternative for Assessing Goodness of Fit of Logistic Regression

Emese Verdes[1] and Tamás Rudas[2]

[1] Department of Sociology, Debrecen University, Debrecen, Hungary
[2] Institute of Sociology, Eötvös Loránd University, Budapest, Hungary

Abstract. In this paper the π^* index of fit introduced by Rudas et al. [9] is applied to the model of logistic regression. First, the original definition of π^* is given with its interpretation, then a review is given on logistic regression focusing on how to assess model fit in traditional ways. Assessing fit often requires grouping of the data and the main part of this paper is concerned with methods for grouping the data and choosing computational technics. These are illustrated using a standard set of data.

1 The π^* Index of Fit

The π^* index was introduced by Rudas et al. [9] for contingency table analysis to propose an alternative of the chi squared measures, especially in the cases, where the traditional measures are not appropriate. If P is an observed contingency table and $\mathbf{M}$ is a model then the π^* index is defined by

$$\pi^*(P, \mathbf{M}) = \inf\{\pi : P = (1-\pi)M + \pi R, \ M \in \mathbf{M}, R \in \mathbf{P}, \ 0 \le \pi \le 1\}, \quad (1)$$

where P, M and R are contingency tables of the same size and $\mathbf{P}$ is the set of all contingency tables of this size. Roughly speaking, the goal is to decompose an observed table into two parts, a first, that fits the model exactly, and a second one that is unrestricted, in the best possible way, namely that the sum of the cell entries in the first part will be maximal. The proportion of the sum of the cell entries in the second part is the π^* value. This is the fraction of the population that cannot be described by the model in the best case. Hence if π^* is small, we will conlude that we are close to the model $\mathbf{M}$, as only a small fraction of the population cannot be described by this model. On the contrary, if π^* is big, we will say, that we are not so close to the model $\mathbf{M}$. Note, that P can be both table of probabilities and table of frequencies. This approach can be applied to the whole population or to a sample. Also note, that in the second case we obtain an estimate for the true population parameter π^*. This measure has several advantages over the traditional chi squared based goodness of fit measures. It does not depend on sample size in the sense that multiplying an observed table with a constant the π^* will not change the estimate, it gives a nice impression about the discrepancy between the model and the data and can be extended for any statistical model. Moreover, we can think of P, M and R as probability measures. From this follows, that if p, m and r are density functions on a statistical space $(\Omega, \mathcal{A}, \mathbf{P})$, then the definition of π^* can be reformulated in the sense that the density p can be represented as a mixture of two densities of the form

$$p = (1 - \pi)m + \pi r, \tag{2}$$

where m comes from the model and r is the density of an unrestricted R from $\mathbf{P}$. In statistical models for continuous variables, one usually distinguishes between the test of significance of an effect and the estimate for the size of this effect. Much of the criticism concerning the performance and applicability of chi squared based statistics is related to the fact that they are appropriate as tests of significance and perform very poorly as estimated effect size. The π^* is clearly an effect size and has attractive features as such. As confidence intervals may be constructed for its true value (see Rudas et al. [9]), it may also be used in testing.

2   Logistic Regression

Logistic regression is an increasingly popular statistical method used in many areas, e.g. in the social sciences. Here a binary response variable is related to one or more potential explanatory variables through the so called logistic function

$$\log \frac{P(Y = 1)}{P(Y = 0)} = D(i, \cdot)\theta, \tag{3}$$

where Y is the response variable, $D(i, \cdot)$ is the i-th row of the design matrix (the i-th setting of the explanatory variables) and θ is the vector of the model parameters. θ is estimated by the ML method. However, evaluating goodness of fit is not so easy. There are different methods proposed. When the number of distinct covariate vectors is relatively small comparing to the sample size n, the traditional chi squared method (Agresti [1]) can be applied. Difficulties arise with continuous covariates where the number of distinct covariate vectors is close to n. In these cases, very often the observations are grouped using some grouping strategy. The most popular test, that is used by most of the computer packages is Hosmer and Lemeshow's test [5]. They group the observations according to the predicted probabilities of the event putting approximately the same number of subjects in each group and then compare the expected and observed frequencies using the chi squared statistic. Problems arise when the estimated probabilities approach either zero or one which is the case in many applications due to the above grouping strategy. Another problem is that different computing packages form different groups and altough all of them apply Hosmer and Lemeshow's test, they conclude to different results [7]. Another possibility is to compute a measure in the spirit of R^2 of ordinary least squares regression. A traditional way of it to compute the proportion of cases predicted correctly. Let the predicted value of the response variable be 1 if the predicted probability of the event is greater than 0.5 and let it be 0 otherwise. This measure has several problems (Weisberg [13]). In particular, there is no baseline or null expectation to compare the correct prediction rate with. Other measures of this type, called pseudo R^2 measures are outlined in Aldrich and Nelson [2] and McKelvey and Zavoina [6]. The backdraw of these measures is, that they are based on the assumption that a dichotomous dependent variable is only a proxy for the true interval level dependent variable that cannot be measured properly and whenever the dependent variable is truly binary, this assumption is not valid. Our π^* approach belongs to the first group of indices. First an appropriate grouping strategy will be chosen based on the theory of multivariate histograms and then the π^* index will be computed using these groups.

3 The Data

As a numerical example, we consider Finney's data [3] used in many textbooks to illustrate logistic regression. The data consist of 39 observations with two covariables. The response is the occurence of restriction on the skin of the digits, and the covariables are the rate and volume of inspired air. After fitting a logistic regression model to the data we have the following results indicating that both covariables are significant.

Table 1. Logistic regression results

θ	SE	Wald	Sig.
−25.89	9.32	7.71	0.005
12.12	4.33	7.81	0.005
10.79	4.19	6.63	0.001

Assessing goodness of fit the Hosmer-Lemeshow test gives different results using different statistical packages as shown in Table 2. Altough all the five software packages are performing the same goodness of fit test, they are obviously using different algorithms to form the groups, which results in radically different conclusion about the goodness of fit.

Pigeon and Heyse [7] reanalised these data. They formed only 4 groups of the observations (instead of 10 formed by the above statistical packages) as they found the number of observations were too small for more groups. The test statistic they used was a modification of the Pearson chi squared

Table 2. Results of the Hosmer-Lemeshow test using different computer packages

computer package	number of groups	χ^2	df	Sig.
SAS	10	24.23	8	0.002
Minitab	10	7.81	8	0.453
SPSS	10	11.10	8	0.195
BMDP	10	17.25	8	0.028
SYSTAT	10	20.92	8	0.007

statistic:

$$J^2 = \sum_{i=1}^{2}\sum_{j=1}^{g} \frac{(O(i,j) - E(i,j))^2}{\phi(j)E(i,j)} \tag{4}$$

where $\phi(j)$ is an adjustment factor handling the underdispersion in the chi squared distribution, $O(i,j)$ and $E(i,j)$ are the observed and expected frequencies for the events and the nonevents in the g groups. Pigeon and Heyse has proved [8] that this statistic has an asymptotic chi squared distribution with $g - 1$ degrees of freedom. Their grouping strategy was also different, they grouped the data according to a chosen covariable. The authors argued that modifications were needed as in the Hosmer-Lemeshow test very often the estimated probabilities approach either 0 or 1 for the first and the last groups according to the grouping strategy putting the low probabilities and high probabilities for events together and so the chi squared test has failed. Pigeon and Heyse's results sorting and grouping the observations by the first or the second covariable can be found in Table 3. As no significant lack of fit could be detected under any of their two grouping strategies the authors concluded that the model provided a reasonable fit of the data.

Table 3. Results of the test proposed by Pigeon and Heyse

covariable used for grouping	number of groups	J^2	df	Sig.
$X1$	4	0.49	3	0.920
$X2$	4	3.28	3	0.350

4 The π^* Index of Fit for Logistic Regression

The question is how to divide the observations into two parts, a first one that fits the model exactly and a second, unrestricted part optimally, i.e. so, that the number of the observations in the first part will be maximal. In contingency tables a possible way is the minimax algorithm proposed by Verdes [12]. When the observations are grouped in logistic regression, the problem is very similar as we have a $2 \times g$ two-way table with observed frequencies given by the data, and estimated frequencies expressed by the model parameters using (3). So everything is given to prepare the functions of the minimax algorithm. Based on a theorem in [10] these functions are the ratios of the observed and expected frequencies in the $2 \times g$ cells depending on the model parameters θ (Verdes [12]). The problem now is to form groups. Choosing different grouping strategies and forming different numbers of groups affects the π^* value as well as it did the chi squared statistic. Then, what is a good grouping strategy, or putting it another way, what is a good estimate of the empirical distribution? The answer will be based on the theory of multivariate histograms (Scott [11]). There are different rules and suggestions about how many groups to choose for a multivariate histogram. Most of these rules suggest to make three groups by each covariable in our data. Preparing now the histograms for the observations $Y = 1$ (and similarly $Y = 0$) we see that this is a very rough approximation of the empirical density, moreover there are bins where the observed frequency is 0 and that causes problems also in the π^* theory. (Fig. 1)

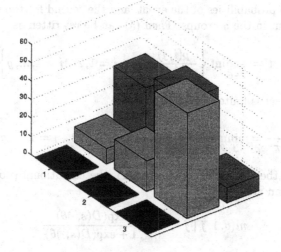

Fig. 1. Histogram for Finney's data $(Y = 1, g = 3 \times 3)$

A smoothing is possible by preparing the so called Averaged Shifted Histogram (ASH) based on the above histogram. The idea of this is to divide each bin into m parts and choose the value of the histogram depending not only on the number of the counts in the small bins, but with decreasing weight, on the counts in the near and further small neighbouring bins. It can be shown (Scott [11]) that as $m \to \infty$, the limiting ASH is the kernel density estimator with the kernel function of the isosceles triangle density. Thus the ASH provides a direct link to the well known kernel methods. As kernel estimators are usually slow to compute, the ASH is a natural candidate for computation.

4.1   The Minimax Algorithm

The minimax algorithm for computing the π^* index is based on the following theorem of Rudas [10]:

Theorem 1. *For the densities $m(\theta)$ and p defined in (2)*

$$1 - \pi^* = \sup_{\theta \in \Theta} \inf_{supp\, m(\theta)} \frac{p}{m(\theta)}, \tag{5}$$

where $supp\, m(\theta)$ stands for the support of $m(\theta)$, and θ is the vector of model parameters.

Having done g groups of the observations, the above support consists of $2g$ points. p and $m(\theta)$ will be the observed and estimated probabilities in these discrete points that we can consider as cells of a $2 \times g$ table, where the first row stands for the probabilities of the event, and the second for the probabilities of the nonevent in the g groups. Then (5) can be rewritten as

$$1 - \pi^* = \sup_{\theta \in \Theta} \inf_{i,j} \left\{ \frac{p(i,j)}{m(\theta,i,j)}; \quad i = 1,2 \quad j = 1,...,g \right\}. \tag{6}$$

As the above set is finite, one can also write

$$\frac{1}{1 - \pi^*} = \min_{i,j} \max_{\theta \in \Theta} \left\{ \frac{m(\theta,i,j)}{p(i,j)}; \quad i = 1,2 \quad j = 1,...,g \right\}. \tag{7}$$

According to the logistic regression model, the conditional probabilities in the s-th column are

$$m(\theta,1,j \mid j = s) = \frac{\exp(D(s,\cdot)\theta)}{1 + \exp(D(s,\cdot)\theta)} \tag{8}$$

$$m(\theta,2,j \mid j = s) = \frac{1}{1 + \exp(D(s,\cdot)\theta)} \tag{9}$$

where $D(s,\cdot)$ denotes the s-th row of the design matrix and θ is the vector of the model parameters. The probability of falling in the s-th group can be estimated from the sample. Denoting by $n(1,s)$ the observed frequency of the event and by $n(2,s)$ the observed frequency of the nonevent in the s-th group, it is $(n(1,s)+n(2,s))/n$. Substituting these expressions into (7) we get

$$\frac{1}{1-\pi^*} = \min_{\theta \in \Theta} \max_s \left\{ \frac{\frac{\exp(D(s,\cdot)\theta)}{1+\exp(D(s,\cdot)\theta)}}{\frac{n(1,s)}{n(1,s)+n(2,s)}}, \frac{\frac{1}{1+\exp(D(s,\cdot)\theta)}}{\frac{n(2,s)}{n(1,s)+n(2,s)}}; \quad s = 1,\ldots,g \right\} \quad (10)$$

which is a minimax problem that can be solved by the MATLAB package.

4.2   Averaged Shifted Histograms (ASHs)

In the univariate case, ASHs are constructed in the following way. Consider a collection of m histograms, $\widehat{f}_1, \widehat{f}_2 ..., \widehat{f}_m$, each with bin width h, but with bin origins $0, \frac{h}{m}, ..., \frac{(m-1)h}{m}$, respectively. The naive or unweighted averaged shifted histogram is defined as

$$\widehat{f}(\cdot) = \frac{1}{m}\sum_{i=1}^{m} f(\cdot). \quad (11)$$

Multivariate ASHs are constructed by averaging shifted multivariate histograms, each with bin width $h_1 \times h_2 \times ... \times h_d$. Then, the multivariate ASH is the average of $m_1 \cdot m_2 \cdot ... \cdot m_d$ shifted histograms shifted by the d coordinate axes all possible ways. In the bivariate case the ASH is given by

$$\widehat{f}(\cdot,\cdot) = \frac{1}{m_1 m_2}\sum_{i=1}^{m_1}\sum_{j=1}^{m_2} f_{ij}(\cdot,\cdot). \quad (12)$$

For a univariate ASH, let (a,b) denote the interval the observations fall into, and let g denote the number of bins with length h in the above interval. Dividing each bin into m equal parts, one can obtain mg bins. Denote them by B_l and let v_l be the bin count in B_l, $l = 1,...,mg$. Define further $v_j = 0$ for $j < 1$ and $j > mg$. The height of the ASH in B_l is the average of the heights of the m shifted histograms, each of width h:

$$\frac{v_{l+1-m} + ... + v_l}{nh}, \quad \frac{v_{l+2-m} + ... + v_{l+1}}{nh}, \quad ..., \quad \frac{v_l + ... + v_{l+m-1}}{nh}. \quad (13)$$

Hence, a general expression for the naive ASH is

$$\widehat{f}(x;m) = \frac{1}{m}\sum_{i=1-m}^{m-1} \frac{(m - |i|)v_{l+i}}{nh} \quad (14)$$

$$= \frac{1}{nh}\sum_{i=1-m}^{m-1} (1 - \frac{|i|}{m})v_{l+i}, \quad x \in B_l.$$

The weights on the bin counts in (14) take on the shape of an isosceles triangle with base $(-1,1)$. However, other weights are also possible. The general ASH uses arbitrary weights $w_m(i)$ and is defined by

$$\widehat{f}(x;m) = \frac{1}{nh} \sum_{i=1-m}^{m-1} w_m(i)\nu_{l+i}, \qquad x \in B_l. \tag{15}$$

In order that $\int \widehat{f}(x;m)dx = 1$, the weights must sum to m in (15). An easy way to define the general weights is

$$w_m(i) = m \times \frac{K(i/m)}{\sum_{i=1-m}^{m-1}K(i/m)}, \qquad i = 1-m,...,m-1, \tag{16}$$

where K is a continuous function defined on $(-1,1)$. K is often chosen to be a probability density function, such as

$$K(t) = \frac{15}{16}(1-t^2)^2_+ = \frac{15}{16}(1-t^2)^2 I_{[-1,1]}(t), \tag{17}$$

which is called the biweight kernel or quartic kernel. Hence an algorithm for the generalized ASH can be given the following way. Step 1. Construct an equally spaced mesh of width $\delta = h/m$ over the interval (a,b), and compute the corresponding bin counts $\{v_l, \quad l = 1,...,mg\}$ for the n data points. Typically, $\delta \ll h$. Step 2. Compute the weight vector, $\{w_m(i)\}$, as in (16). Step 3. Compute $\{f_l, \quad l = 1,...,mg\}$. This can be done in an efficient manner reordering the operations in (15). Rather than computing the ASH estimates individually in each bin, a single pass is made through the bin counts, and the 'effects' of the bins on $f_l, \quad l = 1,...,mg$ are computed. This modification avoids repeated weighting of empty bins. Note that the algorithm for the univariate ASH can be easily extended to the multivariate case, the only difference is that the parameters in the univariate ASH become vectors. In the above algorithm, the precise choice of m is unimportant as long as it is greater than 2 and h is well chosen (Scott [11]). However, many authors studied the limiting behavior of the ASH as $m \to \infty$. It can be showed, that the limiting ASH can be written as

$$\widehat{f}(x) = \frac{1}{nh}\sum_{i=1}^{n}K\left(\frac{x-x_i}{h}\right), \tag{18}$$

where x_i is the i-th data point and $K(\cdot)$ is the kernel function of the isosceles triangle density defined by

$$K(t) = (1-|t|)I_{[-1,1]}(t). \tag{19}$$

The estimator (18) is called the general kernel density estimator with kernel K, corresponding to the generalized ASH defined in (15).

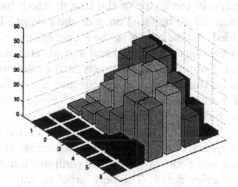

Fig. 2. ASH for Finney's data ($Y = 1$, $g = 3 \times 3$, $m = 2$)

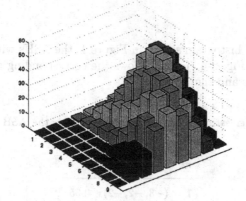

Fig. 3. ASH for Finney's data ($Y = 1$, $g = 3 \times 3$, $m = 3$)

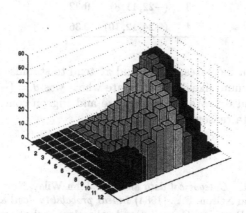

Fig. 4. ASH for Finney's data ($Y = 1$, $g = 3 \times 3$, $m = 4$)

Graphically, what happens is that the kernel estimate places a probability mass of size $1/n$ in the shape of the kernel, which has been scaled by the smoothing parameter h, centered on each data point. These probability masses are then added vertically to give the kernel estimate. So the arbitrary grouping of the observations are avoided and a good estimate of the empirical density can be given. Note, that other kernels such as the normal density can be also used. Based on this connection of ASH and kernel density estimators, our computational method for computing the π^* index of fit is a reasonable choice, especially as kernel estimators are notoriously slow to compute (Scott [11]). The ASH is a 'bona fide' density estimator and a natural candidate to computation. It is a special case of a more general framework called WARPing developed by Hardle and Scott [4] where the computational efficiency of the ASH is discussed in more detail. The only problem can be dimensionality, as for dimensions more than 4, it is generally not possible to fit arrays of sufficient dimension directly in computer memory.

5   Results

Starting from the histogram shown in Figure 1, the ASHs with $m = 2, 3$ and 4 can be seen on Fig. 2-4. Table 4 shows the corresponding estimates of the model parameters and the π^* values.

Table 4. Parameter estimates and π^* values based on the ASH with $m = 1, 2, 3$ and 4

m	θ	π^*
1	$(-3, -2, -2)$	0.55
2	$(-13, 7, 4)$	0.3
3	$(-22, 11, 8)$	0.32
4	$(-24, 12, 10)$	0.36

Note that as m increases, both θ and π^* tend to stabilise and θ is getting close to the maximum likelihood estimate which was $\theta = (-25, 12, 10)$. The π^* value is around 0.3. We can join Heyse and Pigeon concluding that this model fits the data reasonably well.

References

1. Agresti, A. (1990) *Categorical data analysis.* John Wiley, New York
2. Aldrich, J.H. and Nelson, F.E. (1984) *Linear probability, logit and probit models.* Sage University Papers on Quantitative Applications in the Social Sciences. Sage, Beverly Hills

3. Finney, D.J. (1947) The estimation from individual records of the relationship between dose and quantal response. Biometrika **34**, 320-334
4. Hardle, W. and Scott, D.W. (1998) Smoothing in low and high dimensions by weighted averaging using rounded points. Technical report, 88-116, Rice University
5. Hosmer, D.W. and Lemeshow, S. (1980) A goodness-of-fit test for the multiple logistic regression model. Communications in Statistics A**10**, 1043-1069
6. McKelvey, R.D. and Zavoina W. (1976) A statistical model for the ordinal level dependent variables. Journal of Mathematical Sociology **4**,103-120
7. Pigeon, J.G. and Heyse, J.F. (1999) A cautionary note about assessing the fit of logistic regression models. Journal of Applied Statistics **26 7**, 847-853
8. Pigeon, J.G. and Heyse, J.F. (1999) An improved goodness of fit statistic for probability prediction models. Biometrical Journal **41**, 71-82
9. Rudas, T., Clogg, C.C., Lindsay, B.G. (1994) A new index of fit based on mixture methods for the analysis of contingency tables. Journal of Royal Statistical Society Ser. B **56**, 623-639
10. Rudas, T. (1999) The mixture index of fit and minimax regression. Metrika **50**, 163-172
11. Scott, D.W. (1992) *Multivariate density estimation. Theory, practice and visualization.* Wiley and Sons, New York
12. Verdes, E. (2001) MATLAB Code to Compute the π^* Index for the Examples of the Chapter 12 'A latent class approach to measuring the fit of a statistical model' in Hagenaars, McCutcheon (eds.) : *Applied Latent Class Analysis.* Cambridge University Press
13. Weisberg, H.W. (1978) Evaluating theories of congressional roll-call voting. American Journal of Political Science **22**, 554-557

Part IV

Statistics of Stationary Processes

Consistent Estimation of Early and Frequent Change Points

Michael I. Baron[1] and Nira Granott[2]

[1] Department of Mathematics, University of Texas at Dallas, Richardson, USA
[2] Eliot-Pearson Department of Child Development, Tufts University, Medford, USA

Abstract. We address two types of processes with change points that often arise in practical situations. These are processes with *early change points* and processes with *frequent change points*. Early change points may occur after very few observations and may be followed by additional change points or more complicated patterns. Frequent change points separate different homogeneous phases of the observed process with the possibility of very short phases.

Uncertainty of the considered processes during their later phases forces the use of sequential tools, in order to minimize samples from later phases. Change-point detection and post-estimation schemes for these situations are developed. They possess a number of desired properties, not satisfied by procedures proposed in the earlier literature. One of them is distribution-consistency. Unlike the traditional concept of consistency, it implies convergence of small-sample change-point estimators to the corresponding parameters as the magnitude of changes tends to infinity.

1 Introduction and Examples

In classical change-point problems, the distribution of observed data changes at an unknown moment, which is the parameter of interest. Sample $X = (X_1, X_2)$ consists of two subsamples from distributions f and g, respectively, separated by an unknown change point ν. A vast amount of literature covers sequential and retrospective methods of change-point estimation. Classical references are eg. [7], [8], [9], [10], [14], [15], [23], [24], [26]; also see [5], [6], [16], and [29] for a survey of on-line and off-line algorithms.

Most of the proposed estimation schemes assume exactly one change point in the observed process and sufficiently large pre-change and post-change sample sizes. However, both assumptions are violated in a number of applications including developmental, cognitive, and educational psychology, energy pricing, meteorology, and quality control.

In problem solving processes, the first insight about a solution can occur after only a few solution attempts. Consequently, the first change point in the distribution of solution times represents the moment of the first discovery and the end of a trial-and-error phase. It may occur after very few solutions, leading to a problem of change-point detection from small data sets ([3], see Fig. 1).

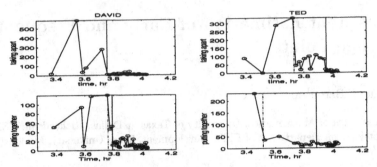

Fig. 1. Early change points in problem solving processes. Two participants in an experiment, David and Ted, repeatedly take twisted nails apart and put them together until they know the solution thoroughly. Solution times in each task (in sec) are plotted against the experiment's time scale (in hrs). The first discovery occurs after very few solutions and yields significant reduction of the solution times. After that, the distribution of David's solution times stabilizes, whereas Ted has at least one additional change point and a further reduction of solution times.

Typical processes of learning and development consist of several phases, separated by change points. A phase can be short (e.g., fast learning in [2]). Such a process is often associated with frequent changes of strategies ([2]). Fig. 2 presents results of an experiment where participants had to learn the functions of a given robot. Different strategies included observing the robot, turning it on and observing its performance, dismantling the robot and exploring its parts. Participants of the study used different strategies subsequently. The frequency of the alternation of strategies changed repeatedly, indicating different phases in the learning process. The times of a continuous use of the same strategy are depicted on Fig. 2. Periods of short single-strategy use times indicate phases of frequent alternation of strategies.

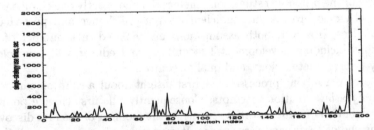

Fig. 2. Multiple, possibly frequent change points in a learning process. The single-strategy use times (sec) are plotted against the number of a strategy switch.

We define a process with an *early change point* to be a sequence of random variables $\mathbf{X} = (\mathbf{X}_1, \mathbf{X}_2, \dots)$, where

$$\begin{cases} \mathbf{X}_1 = (X_1, \dots, X_\nu) & \sim f, \quad \nu \geq 1 \\ \mathbf{X}_2 = (X_{\nu+1}, \dots, X_\kappa) \sim g, \quad \kappa \geq \nu + 1 \\ \mathbf{X}_3, \mathbf{X}_4, \dots, \dots & \sim ? \end{cases} \tag{1}$$

After the first change point ν, observations come from distribution $g \neq f$ until some unknown moment κ. After κ, no tractable model is assumed, as the later part of the process may contain further change points or even more complicated patterns and trends. The objective is to estimate the first change point ν.

A process with *multiple change points* will then be described as $\mathbf{X} = (\mathbf{X}_1, \mathbf{X}_2, \mathbf{X}_3, \dots, \mathbf{X}_\lambda)$, where

$$\begin{cases} \mathbf{X}_1 = (X_{\nu_0}, \dots, X_{\nu_1}) & \sim f_1 \\ \mathbf{X}_2 = (X_{\nu_1+1}, \dots, X_{\nu_2}) & \sim f_2 \\ \dots\dots\dots\dots\dots\dots \\ \mathbf{X}_\lambda = (X_{\nu_{\lambda-1}+1}, \dots, X_{\nu_\lambda}) \sim f_\lambda \end{cases} \tag{2}$$

where $f_1, \dots, f_\lambda$ are either known or unknown densities $(f_i \neq f_{i+1})$, $\nu_1, \dots, \nu_{\lambda-1}$ are change points (with a convention that $\nu_\lambda = N$ is the total sample size and $\nu_0 = 1$), and λ is the unknown number of homogeneous subsamples. If $|\nu_{k+1} - \nu_k|$ is small for some k, we will call this model a process with *frequent change points*.

Notice that densities f_i and f_j may coincide if $|i - j| \geq 2$. In the extreme case, all the odd subsamples come from the same distribution f_1, and all the even subsamples come from the same distribution f_2, so that the process oscillates between two densities f_1 and f_2. Such a model is suitable for detrended electricity prices, where relatively long "regular" or "control" periods alternate with short-term "spikes" ([4], [12], see Fig. 3).

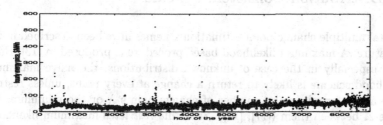

Fig. 3. Time plot of energy prices in New England (from [11]). Detrended and deseasonalized prices have multiple change points, representing first and last hours of "spikes". The process shifts from the *regular* mode to the *spike mode* and vice versa, and it is described by just two distributions.

The objective here is to estimate $(\lambda; \nu_1, \ldots, \nu_{\lambda-1})$, a parameter of an unknown dimension. The possibility of a homogeneous sample with no change points ($\lambda = 1$) is not excluded.

In both (1) and (2), the search for the first change point is complicated by uncertain behavior of the process after the end of the second phase. Thus, although the problems of estimating early or frequent change points are *retrospective* (not sequential) in general, it is not recommended to use retrospective statistical procedures, because they utilize the entire data set including the uncertain phases of the process. Conversely, it is desirable to use *sequential* tools that will utilize the minimum number of observations and stop as soon as the change point is detected.

Essentially, for the process with an early change point, one can resample one observation at a time until the first change is detected. In a situation with multiple change points, one will estimate the detected change point, discard all the pre-change observations, and search for the next change point sequentially. Therefore, both types of estimation problems, early change points and frequent change points, can be treated by similar techniques based on sequential detection and post-estimation of change points.

Many competing sequential schemes can be proposed. We select an optimal procedure according to the introduced principle of *distribution consistency*. It implies convergence of each change-point estimator to the corresponding parameter when sample sizes are fixed, but the magnitude of a change becomes more and more significant. In problems with early and frequent change points yielding short phases, we find this property at least as important as consistency in the classical sense.

The algorithm and its optimal properties are described in Section 2. In Section 3, we use the proposed scheme to detect possible global climate changes, spikes of electricity prices, and different phases in learning and problem solving processes.

2   Sequential Estimation of Multiple Change Points: A Distribution Consistent Scheme

Several multiple change-point estimation schemes have been described in the literature. A maximum likelihood based procedure is proposed in [13]. However, especially in the case of unknown distributions, the naive maximum likelihood scheme is likely to return a change at every point, unless restrictions are enforced on the number of change points λ or the minimum distance Δ between them ([17], [18]). Still, this *restricted* maximum likelihood scheme tends to detect too many change points. For example, in the case of Bernoulli(p) observations, it will detect change points in any sample of size $N > 2\Delta$ if and only if at least two observations are different ([3]). Thus, the probability of a false alarm is as high as $1 - p^N - (1 - p)^N$ in this case.

A conceptually different *binary segmentation* scheme ([28]) is an iterative procedure that divides the observed sample into two most distant subsamples, then divides each subsample, etc., until all the obtained subsamples are homogeneous. The disadvantage of this scheme is that no more than one change point is assumed at each step. For example, in the case of two intermittent distributions, as on Fig. 3, it is unlikely to find a point separating two significantly different subsamples.

These problems can be resolved by a sequential estimation scheme that (1) considers increasing subsamples instead of the entire sample that may contain complicated patterns; (2) detects one change point at a time and does not assume its uniqueness in the observed data; (3) has an option of detecting 0 change points; (4) is sufficiently sensitive in order to detect a change point occurring after a short phase.

For the process with an early change point, the scheme consists of several steps outlined below. For the process with frequent change points, these steps are repeated until the entire data set is resampled.

2.1 Step 1: Sequential Detection

Observations are sampled sequentially until a stopping rule detects a change point. For the examples in Section 1, we used a stopping rule

$$T(h) = \inf \{n : W_n \geq h\},$$

based on a *cusum process*

$$W_n = \max_k \sum_{j=k+1}^{n} \log \frac{g(X_j)}{f(X_j)},$$

where f and g are the pre- and post-change densities or probability mass functions. Optimality of $T(h)$ is shown in [19] and [25]. In the case of unknown densities, one uses their best estimates for each "potential" value k of a change point, computes the *generalized likelihood ratio* based cumulative sums

$$\tilde{W}_n = \max_k \sum_{j=k+1}^{n} \log \frac{\hat{g}(X_j)}{\hat{f}(X_j)}, \tag{3}$$

and defines the stopping rule $\tilde{T}(h)$ similarly to $T(h)$. This stopping rule achieves asymptotically equivalent mean delay and mean time between false alarms ([1]).

Facing a possibility of early or frequent change points, one should increase sensitivity of the algorithm by choosing a low threshold h or a high probability of type I error α. The price to be paid is the increasing rate of false alarms, however, false change points will (hopefully) be eliminated at Step 3.

If only a sample of size N is available, all abovementioned stopping rules are curtailed so that $P\{T \leq N\} = 1$. In the case when $T(h) = N$ and $\tilde{W}_n < h$, the scheme results in zero detected change points. In all the other cases, a change point is detected and its location needs to be estimated.

2.2   Step 2: Post-Estimation

Notice that the stopping rule T itself is a poor estimator of the change point ν. Indeed, if $T \leq \nu$, it is a false alarm. If $T > \nu$, it is a biased estimator that always overestimates the parameter. Therefore, the detected change point has to be *post-estimated*, i.e., estimated after its occurrence is detected by a stopping rule.

One way of obtaining an approximately unbiased estimator of ν is to estimate the bias of $T(h)$ and subtract it from $T(h)$. According to [1], this bias, also known as *mean delay*, is asymptotically $(h+C)/K(f,g)$, as $h \to \infty$, where $K(f,g)$ is the Kullback information number, and C is independent of h. In the case of sufficiently long phases before and after the change point, subtracting the estimated bias from $T(h)$ yields an approximately unbiased estimator of ν. However, in the case of frequent change points and unknown densities, no reliable estimators of C and K are available.

A *last-zero* estimator

$$\hat{\nu}_{LZ} = \sup \{k < T(h), \ W_k = 0\},$$

proposed in [21] and [27], is essentially the maximum likelihood estimator of ν, assuming a fixed-size sample rather than a sample of a random size T, which is the stopping rule. The corresponding estimator in the case of unknown densities is

$$\tilde{\nu} = \sup \left\{ k < \tilde{T}(h), \ \tilde{W}_k = 0 \right\}.$$

It can be shown that this estimator fails to satisfy an important property of *distribution consistency* ([3]).

Definition 1. *Consider a family of distributions $\mathcal{F}$ and a nonnegative function $\mathcal{D}$ on $\mathcal{F} \times \mathcal{F}$ with*

$$\mathcal{M} = \sup\{\mathcal{D}(f,g)|f,g \in \mathcal{F}\} \leq \infty.$$

Let $\mathbf{X}$ be a fixed-size sample generated according to the multiple change-point model (2) with $\lambda \geq 1$ and $f_j \in \mathcal{F}$ for $j = 1,\ldots,\lambda$. Let $(\hat{\lambda}; \hat{\nu}_1,\ldots,\hat{\nu}_{\hat{\lambda}})$ be an estimator of $(\lambda; \nu_1,\ldots,\nu_\lambda)$. A change-point estimation scheme will be called **distribution consistent** *(with respect to $\mathcal{D}$) if*

$$P(\hat{\lambda} = \lambda) \to 1$$

and

$$P\left(\max\left\{|\hat{\nu}_j - \nu_j|, 1 \leq j < \min(\lambda, \hat{\lambda})\right\} = 0\right) \to 1,$$

as

$$\min_{1 \le j < \lambda} \mathcal{D}(f_j, f_{j+1}) \to \mathcal{M}.$$

This property means that all the change-point estimators converge to the corresponding parameters, as changes become more and more significant but the sample size and all the change points remain fixed. It is implied that function $\mathcal{D}$ measures discrepancy between two distributions. Therefore, a distribution-consistent scheme estimates change points very accurately when any two consecutive phases are generated by significantly different distributions.

This property of a multiple change-point estimation scheme is desirable in all the examples of Section 1, where short phases are followed by entirely different patterns. Certainly, if the magnitude of a change is very significant, one would like to estimate the time of change very accurately, even from small samples.

Notice that for integer-valued λ and $\{\nu_j\}$, distribution consistency is equivalent to convergence in probability. It also implies that a sample with no change points ($\lambda = 0$) provides no false alarms with the probability converging to 1, and the probability of not detecting a change point in a sample with change points converges to 0.

The $\tilde{W}$-based last-zero estimator $\hat{\nu}_{LZ}$ is not distribution consistent. Indeed, any time when the last zero of $\tilde{W}$ occurs before the true change point ν, the estimator $\hat{\nu}_{LZ}$ is based on a sample from the pre-change distribution f_1 only. If f_1 remains fixed while f_2 drifts away from it so that $\mathcal{D}(f_1, f_2) \to \mathcal{M}$, the distribution of $\hat{\nu}_{NZ}$ does not change and $\hat{\nu}_{NZ}$ does not converge to ν.

For example, consider a change in the parameter of Exponential distribution from θ_0 to θ_1. In this case, the nuisance parameters are estimated by the respective sample means, and according to (3),

$$\hat{W}_n = \max_{k \le n} \{(n - k)(\xi_{kn} - \log \xi_{kn} - 1)\},$$

where $\xi_{kn} = \bar{x}_{kn}/\bar{x}_{0k}$ and $\bar{x}_{ij} = \sum_{i+1}^{j} x_t/(j - i)$. As $\theta_1 \downarrow 0$ or $\theta_1 \uparrow \infty$ while θ_0 remains constant, $\xi_{\nu,\nu+1} - \log \xi_{\nu,\nu+1} - 1 \to \infty$ in probability. Hence,

$$P(T(h) > \nu + 1) \le P(\hat{W}_{\nu+1} \le h) \to 0, \tag{4}$$

that is, the change at ν will be detected no later than at $(\nu + 1)$ with probability converging to 1. Therefore, for any $\epsilon < 1$,

$$P(|\hat{\nu}_{LZ} - \nu| < \epsilon) = P(\hat{\nu}_{LZ} = \nu) = P(\hat{W} = 0 \cap T(h) = \nu + 1) + o(1)$$

$$\to P_{\theta_0} \left(\max_{k \le \nu}(\nu - k)(\xi_{k\nu} - \log \xi_{k\nu} - 1) = 0 \right) < 1,$$

as $\theta_1 \downarrow 0$ or $\theta_1 \uparrow \infty$. Thus, $\hat{\nu}_{LZ}$ (in presence of nuisance parameters) is not distribution consistent.

However, distribution-consistent schemes exist. One of them is based on the cusum stopping rule $\tilde{T}$ and the *minimum p-value* estimator

$$\hat{\nu}_{MP} = \arg \min_{1 \leq k < \tilde{T}} p(k, \tilde{T}, \mathbf{X}),$$

where $p(k, \tilde{T}, \mathbf{X})$ is the p-value of the likelihood ratio test comparing subsamples $\mathbf{X_1} = (X_1, \ldots, X_k)$ and $\mathbf{X_2} = (X_{k+1}, \ldots, X_{\tilde{T}})$.

2.3   Step 3: Tests of Significance

To eliminate false alarms, significance of each detected change point has to be tested. Likelihood ratio tests are easy to implement here, and significance of the detected change point is measured by the minimum p-value $p(\hat{\nu}_{MP}, \tilde{T}, \mathbf{X})$.

If the test is significant, one applies steps 1–3 to the post-change subsample $\{X_k, k > \hat{\nu}_{MP}\}$, searching for the next change point. Otherwise, we have a false alarm, and the search continues based on the initial sample, or a part of it starting after the last change point that was found significant.

2.4   The Case of Gamma Distributions

Gamma family is a suitable model for solution times and single-strategy use times (Fig. 1 and Fig. 2), see [3] for the results of goodness-of-fit tests.

The assumption of independence of solution times is justified by the following nonparametric test that was applied to the processes of problem solving. For each of the participants who had at least two solutions, we counted the number of pairs (X_i, X_{i+1}) of consecutive solution times that are on one side of their sample median m, and the number of pairs that are on different sides. If solution times are independent in each pair, then $X_{i+1} > m$ with probability 0.5, independently of X_i. However, if X_i and X_{i+1} are positively (negatively) dependent, the probability of $(X_i - m)(X_{i+1} - m) > 0$ is greater (smaller) than 0.5. This sign test that is expressed as a simple test about the population proportion did not reject the hypothesis of independence (against a short-term dependence) with a p-value of 0.13.

For simplicity, consider a family of Gamma distributions with the same known shape parameter α and unknown scale parameter β that changes at every change point. For any two members of this family, it is natural to consider the discrepancy function

$$\mathcal{D}(f_\beta, f_{\beta^*}) = \max\{\beta/\beta^*, \beta^*/\beta\}.$$

For each $k \leq \tilde{T}$, the p-value is computed as

$$p(k, \tilde{T}, R) = 1 - \left\{ \bar{B}\left(\frac{b}{b+1}, m, n\right) - \bar{B}\left(\frac{a}{a+1}, m, n\right) \right\}$$

where a and b are the only two positive roots of the equation $\Lambda(R) = \Lambda_{observed}$ for the likelihood ratio test statistic $\Lambda(R)$,

$$R = \sum_{j \le \hat{\nu}_{MP}} X_j \left(\sum_{j > \hat{\nu}_{MP}} X_j \right)^{-1},$$

and $\bar{B}$ denotes the incomplete Beta function. The statistic is computed as

$$\Lambda(R) = \left(\frac{k^k (\tilde{T} - k)^{\tilde{T}-k}}{\tilde{T}^{\tilde{T}}} (1 + R^{-1})^k (1 + R)^{\tilde{T}-k} \right)^\alpha$$

and graphed on Fig. 4.

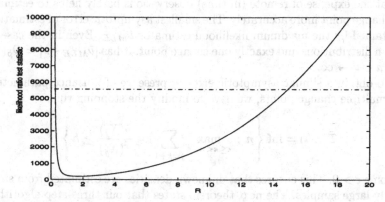

Fig. 4. The likelihood ratio test statistic as a function of R. If the horizontal line represents the observed value of Λ, we have exactly two roots a and b of the equation $\Lambda(R) = \Lambda_{observed}$.

Theorem 1. *(Distribution consistency of the proposed scheme) Suppose that*

(i) *Gamma*(α_j, θ_j) *distribution changes to Gamma*$(\alpha_{j+1}, \theta_{j+1})$ *at a change point ν_j, $j < \lambda$;*

(ii) *α_j are known, θ_j are unknown, and $\rho_j = \theta_{j+1}/\theta_j$;*

(iii) *$h \to \infty$ and $h/\min_j |\log \rho| \to 0$ as $\rho \to 0$.*

Then the following probabilities converge to 1 as $\rho \to 0$:

(a) *the probability of no false alarms, $1 - P(\tilde{T} \le \nu)$,*

(b) *the probability of detecting a change, $P(\tilde{T} < N \mid \nu \le N)$,*

(c) *the probability of a minimal delay, $P(\tilde{T} = \nu + 1)$,*

(d) *the probability that the detected change point is found significant,*
   $P(p(\hat{\nu}_{MP}, \tilde{T}, \mathbf{X}) < \delta)$, for any $\delta > 0$,

(e) the probability of estimating with no error, $P(\hat{\nu}_{MP} = \nu)$.

Corollary 1. *The multiple change-point estimation scheme $(\tilde{T}, \hat{\nu}_{MP})$ is distribution-consistent.*

In fact, Theorem 1 establishes a stronger property than the distribution consistency. That is, detection of each change point requires just one post-change observation with the probability converging to 1. It is a valuable property in any problem where complicated patterns force to use minimum data to detect a change point. For the proof of Theorem 1, see [3].

2.5   Classical Consistency

It is generally known (see [14]) that change-point estimators are not consistent in a classical sense. It agrees with the intuition, since increasing the sample size at the expense of remote (in time) observations barely helps to estimate the change point more accurately. The smallest asymptotic error of estimation is attained by the maximum likelihood estimator $\hat{\nu}_{MLE}$. Even in the case of known distributions and exactly one change point, it has $|\hat{\nu}_{MLE} - \nu| = O_p(1)$, as $\nu, n - \nu \to \infty$.

To obtain a similar asymptotic error in presence of nuisance parameters and multiple change points, we have to modify the stopping rule,

$$\tilde{T}(\epsilon, h) = \inf\left\{ n : \max_{\epsilon \leq k < n - \epsilon} \sum_{j=k+1}^{n} \log \frac{\hat{g}(X_j)}{\hat{f}(X_j)} \geq h \right\}$$

for some $\epsilon > 0$. This ensures that unknown densities are estimated from sufficiently large samples. The next theorem states that our three-step algorithm (with the threshold h being a function of the sample size N) provides the same asymptotic error of multiple change-point estimators. At the same time, the number of change points $\lambda - 1$ is estimated consistently in the classical sense.

Theorem 2. *(Sample consistency) Assume that*

(i) all the densities in (2) belong to a canonical exponential family

$$f_j(x) = f(x|\theta_j) = f(x|0) \exp\{\theta_j x - \psi(\theta_j)\};$$

(ii) there are no fake change points, i.e., $\theta_j = \theta_k$ yields $|j - k| \geq 2$;
(iii) all nuisance parameters are estimated by the method of maximum likelihood;
(iv) sample size N, threshold h, and the smallest distance between change points $\Delta = \min_j \{\nu_{j+1} - \nu_j\}$ satisfy the following conditions,

$$\Delta \to \infty, \quad N \exp\left\{ -\frac{\epsilon(N) h(\epsilon(N))}{N} \right\} \to 0, \quad \frac{h(N)}{N - \epsilon(N)} \to 0, \quad \frac{h(N)}{\Delta(N)} \to 0,$$

as $N \to \infty$.

Then

(a) $\hat{\lambda}$ *is consistent, i.e.,* $P\{\hat{\lambda} = \lambda\} \to 1$, *as* $N \to \infty$, *for any* $\lambda \geq 0$;

(b) $\max_{1 \leq j < \lambda} |\hat{\nu}_j - \nu_j| = O_p(1)$, *as* $N \to \infty$, *where* $\{\hat{\nu}_j\}$ *are change point estimates in their ascending order.*

Conditions of this theorem are trivially satisfied if, say, $\epsilon(N) = \epsilon N$, $\Delta(N) = \Delta N$, and $h(N)$ is any function whose rate of growth is between $\log(N)$ and N. The Theorem is proved by induction in λ, $\lambda = 0, 1, \ldots$, through Chernoff-type inequalities for generalized likelihood ratios in the case of exponential families. For the detailed proof, see [2].

3 Applications and Practical Results

We used the described algorithm to identify global changes of climate, spikes in hourly electricity prices, and different phases in development, learning, and problem solving processes.

Analysis of Central England Temperatures data (see [20],[22]) shows significant changes of climate around 1730, 1830, and in the 1940s. The former two may be attributed to the Little Ice Age period whereas the latter is likely to be related to a greenhouse effect ([1]).

Analysis of detrended electricity prices (Fig. 3) allows to separate spikes and to fit a suitable Markovian model with transitions from the regular state to the spike state and vice versa. Then, an appropriate ARMA model is fit to the transformed interspike prices, whereas spikes are modeled by a compound lognormal distribution ([4]). Such a stochastic model is necessary for the prediction of prices, evaluation of futures and forward options, etc.

Analysis of the solution times and single-strategy use times shows a number of different phases in microdevelopmental processes. Comparison of different phases discovers that some earlier patterns are repeated later. A clustering algorithm is then used to match similar patterns and to identify different "types of behavior" ([2]).

Acknowledgements

We thank Professors Yoel Haitovsky, Udi Makov, and Yaacov Ritov for the very well organized conference and their hospitality in Israel. Research of both authors is supported by the U.S. National Science Foundation, grant # 9818959.

References

1. Baron, M. (2000). Nonparametric adaptive change-point estimation and on-line detection. Sequential Anal. **19**, 1–23

2. Baron, M., Granott, N. (2001). Estimation of frequent change points. In preparation

3. Baron, M., Granott, N. (2001). Small sample change-point analysis with applications to problem solving. Submitted

4. Baron, M., Rosenberg, M., Sidorenko, N. (2001). Electricity pricing: modelling and prediction with automatic spike detection. Energy, Power, and Risk Management, October 2001, 70–73

5. Basseville, M., Nikiforov, I.V. (1993). *Detection of Abrupt Changes: Theory and Application.* PTR Prentice-Hall, New Jersey

6. Bhattacharya, P.K. (1995). Some aspects of change-point analysis. In: *Change-Point Problems*, IMS Lecture Notes - Monograph Series **23**, 28–56, Stanford

7. Brodsky, B.E., Darkhovsky, D.S. (1993). *Nonparametric Methods in Change-Point Problems.* Kluwer Academic Publishers, The Netherlands

8. Carlstein, E. (1988). Nonparametric estimation of a change-point. Ann. Stat. **16**, 188–197

9. Chernoff, H., Zacks, S. (1964). Estimating the current mean of a normal distribution which is subject to changes in time. Ann. Math. Stat. **35**, 999–1028

10. Cobb, G.W. (1978). The problem of the Nile: conditional solution to a change-point problem. Biometrika **62**, 243–251

11. ISO New England (2000). Historical Data Archive, http://www.isone.com/Historical_Data/hist_data.html

12. Ethier, R., Dorris, G. (1999). Don't ignore the spikes. Energy & Power Risk Management, July/August

13. Fu, Y.X., Curnow, R.N. (1990). Maximum likelihood estimation of multiple change points. Biometrika **77**, 563–573

14. Hinkley, D.V. (1970). Inference about the change-point in a sequence of random variables. Biometrika **57**, 1–17

15. Khan, R.A. (1979). A sequential detection procedure and the related cusum procedure. Sankhya Ser A **40**, 146–162

16. Lai, T.L. (1995). Sequential changepoint detection in quality control and dynamical systems. J. Roy. Stat. Soc. B **57**, 613–658

17. Lee, C.B. (1995). Estimating the number of change points in a sequence of independent normal random variables. Stat. Probabil. Lett. **25**, 241–248

18. Lee, C.B. (1996). Nonparametric multiple change-point estimators. Stat. Probabil. Lett. **27**, 295–304

19. Lorden, G. (1971). Procedures for reacting to a change in distribution. Ann. Math. Stat. **42**, 1897–1908

20. Manley, G. (1974). Central England Temperatures: monthly means 1659 to 1973. Q.J.R. Meteorol. Soc. **100**, 389–405

21. Montgomery, D.C. (1997). *Introduction to Statistical Quality Control*, Third Ed., Wiley, New York

22. Parker, D.E., Legg, T.P., Folland, C.K. (1992). A new daily central england temperature series, 1772-1991. Int. J. Climatol. **12**, 317–342

23. Pollak, M. (1985). Optimal detection of a change in distribution. Ann. Stat. **13**, 206–227

24. Pollak, M., Siegmund, D. (1987). Sequential detection of a change in a normal mean when the initial value is unknown. Ann. Stat. **15**, 749–779

25. Ritov, Y. (1990). Decision theoretic optimality of CUSUM procedure. Ann. Stat. **18**, 1464–1469

26. Shiryaev, A.N. (1963). On optimal method in earliest detection problems. Theor. Probab. Appl. **8**, 26–51
27. Srivastava, M.S., Wu, Y. (1999). Quazi-stationary biases of change point and change magnitude estimation after sequential cusum test. Sequential Anal. **18**, 203–216
28. Vostrikova, L.J. (1981). Detecting "disorder" in multidimensional random processes. Sov. Math. Dokl. **24**, 55–59
29. Zacks, S. (1983). Survey of classical and Bayesian approaches to the change-point problem: Fixed sample and sequential procedures in testing and estimation. In: *Recent Advances in Statistics*, 245–269, Academic Press, London

Asymptotic Behaviour of Estimators of the Parameters of Nearly Unstable INAR(1) Models

Márton Ispány[1]*, Gyula Pap[1]* and Martien C.A. van Zuijlen[2]

[1] Institute of Mathematics and Informatics, University of Debrecen, Hungary
[2] Department of Mathematics, University of Nijmegen, The Netherlands

Abstract. A sequence of first–order integer–valued autoregressive type (INAR(1)) processes is investigated, where the autoregressive type coefficients converge to 1. It is shown that the limiting distribution of the joint conditional least squares estimators for this coefficient and for the mean of the innovation is normal. Consequences for sequences of Galton–Watson branching processes with unobservable immigration, where the mean of the offspring distribution converges to 1 (which is the critical value), are discussed.

1 Introduction

In many practical situations one has to deal with non–negative integer–valued time series. Examples of such time series, known as counting processes, arise in several fields of medicine (see, e.g., Cardinal et.al. [5] and Franke and Seligmann [9]). To construct counting processes Al–Osh and Alzaid [1] proposed a particular class of models, the so–called INAR(1) model. Later Al–Osh and Alzaid [2], Du and Li [8] and Latour [12] generalized this model by introducing the INAR(p) and GINAR(p) models. These processes can be considered as discrete analogues of the scalar– and vector–valued AR(p) processes, because their correlation structure is similar.

The present paper deals with so–called nearly unstable INAR(1) models. It is, in fact, a sequence of INAR(1) models where the autoregressive type coefficient α_n is close to one, more precisely, $\alpha_n = 1 - \gamma_n/n$ with $\gamma_n \to \gamma$, where $\gamma \geqslant 0$. This parametrization has been suggested by Chan and Wei [6] for the usual AR(1) model. The main motivation of our investigation comes from econometrics, where the so–called 'unit root problem' plays an important role (see, e.g., the monograph of Tanaka [15]). We considered in [10] the conditional least squares estimate (CLSE) for α_n assuming that the mean μ_ε of the innovation is known. In this paper we do not suppose that μ_ε is known, and we show asymptotic normality of the joint CLSE of α_n and μ_ε.

* This research has been supported by the Hungarian Scientific Research Fund under Grant No. OTKA–T032361/2000 and OTKA–F032060/2000.

To define the INAR(1) model let us recall the definition of the $\alpha \circ$ operator which is due to Steutel and van Harn [14].

Definition 1.1 *Let X be a non–negative integer–valued random variable. Let $(Y_j)_{j \in \mathbb{N}}$ be a sequence of independent and identically distributed (i.i.d.) Bernoulli random variables with mean α. We assume that the sequence $(Y_j)_{j \in \mathbb{N}}$ is independent of X. The non–negative integer–valued random variable $\alpha \circ X$ is defined by*

$$\alpha \circ X := \begin{cases} \sum_{j=1}^{X} Y_j, & X > 0, \\ 0, & X = 0. \end{cases}$$

*The sequence $(Y_j)_{j \in \mathbb{N}}$ is called a **counting sequence**.*

Let $(\varepsilon_k)_{k \in \mathbb{N}}$ is an i.i.d. sequence of non–negative integer–valued random variables with mean μ_ε and variance σ_ε^2. The zero start INAR(1) time series model is defined as

$$X_k = \begin{cases} \alpha \circ X_{k-1} + \varepsilon_k, & k = 1, 2, \ldots, \\ 0, & k = 0, \end{cases}$$

where the counting sequences $(Y_j)_{j \in \mathbb{N}}$ involved in $\alpha \circ X_{k-1}$ for $k = 1, 2, \ldots$ are mutually independent and independent of $(\varepsilon_k)_{k \in \mathbb{N}}$. We suppose that $\mu_\varepsilon > 0$ (otherwise $X_k = 0$ for all $k \in \mathbb{N}$).

It is easy to show (see [10]), that

$$\lim_{k \to \infty} \mathsf{E} X_k = \frac{\mu_\varepsilon}{1 - \alpha}, \qquad \lim_{k \to \infty} \mathsf{Var} X_k = \frac{\sigma_\varepsilon^2 + \alpha \mu_\varepsilon}{1 - \alpha^2}, \qquad \text{for all } \alpha \in [0, 1),$$

and that $\lim_{k \to \infty} \mathsf{E} X_k = \lim_{k \to \infty} \mathsf{Var} X_k = \infty$ if $\alpha = 1$. The case $\alpha \in [0, 1)$ is called *stable* or *asymptotically stationary*, while the case $\alpha = 1$ is called *unstable*.

First we recall the results concerning the estimation of the parameter α in case if the value of μ_ε is supposed to be known. Let $\mathcal{F}_k$ be the σ–algebra generated by the random variables $X_1, \ldots, X_k$. Clearly $\mathsf{E}(X_k \mid \mathcal{F}_{k-1}) = \alpha X_{k-1} + \mu_\varepsilon$, thus the conditional least squares estimator (CLSE) $\widehat{\alpha}$ of α based on the observations $(X_k)_{1 \leqslant k \leqslant n}$ (assuming that μ_ε is known) can be obtained by minimizing the sum of squares

$$\sum_{k=1}^{n} (X_k - \alpha X_{k-1} - \mu_\varepsilon)^2 \tag{1}$$

with respect to α, and it has the form

$$\widehat{\alpha}_n = \frac{\sum_{k=1}^{n} X_{k-1}(X_k - \mu_\varepsilon)}{\sum_{k=1}^{n} (X_{k-1})^2}.$$

In the stable case under the assumption $E\varepsilon_1^3 < +\infty$ we have

$$n^{1/2}(\widehat{\alpha}_n - \alpha) \xrightarrow{\mathcal{D}} \mathcal{N}(0, \sigma_{\alpha,\varepsilon}^2), \qquad \sigma_{\alpha,\varepsilon}^2 = \frac{\alpha(1-\alpha)EZ_0^3 + \sigma_\varepsilon^2 EZ_0^2}{(EZ_0^2)^2},$$

where $(Z_k)_{k\in\mathbb{Z}}$ is a stationary solution of the INAR(1) model

$$Z_k = \alpha \circ Z_{k-1} + \varepsilon_k, \qquad k \in \mathbb{Z},$$

see Klimko and Nelson [11].

Let us consider now a nearly unstable sequence of INAR(1) models

$$X_k^{(n)} = \begin{cases} \alpha_n \circ X_{k-1}^{(n)} + \varepsilon_k^{(n)}, & k = 1, 2, \ldots, \\ 0, & k = 0, \end{cases} \qquad n = 1, 2, \ldots,$$

where the autoregressive type coefficient has the form $\alpha_n = 1 - \gamma_n/n$ with $\gamma_n \to \gamma$ such that $\gamma \geqslant 0$. In [10] the authors have proved that $(\widehat{\alpha}_n)_{n\in\mathbb{N}}$ is asymptotically normal, namely,

$$n^{3/2}(\widehat{\alpha}_n - \alpha_n) \xrightarrow{\mathcal{D}} \mathcal{N}(0, \sigma_{\gamma,\varepsilon}^2).$$

In this case it suffices to assume $E\varepsilon_1^2 < +\infty$. We draw the attention to the normalizing factor $n^{3/2}$, which is different from the stable case.

In the present paper we suppose that both the parameters α and μ_ε are unknown. By minimizing the sum of squares (1) with respect to α and μ_ε, we obtain the joint conditional least squares estimator $(\widetilde{\alpha}_n, \widetilde{\mu}_{\varepsilon,n})$ of the vector $(\alpha_n, \mu_\varepsilon)$ based on the observations $(X_k^{(n)})_{1\leqslant k\leqslant n}$:

$$\widetilde{\alpha}_n = \frac{\sum_{k=1}^n X_{k-1}^{(n)}(X_k^{(n)} - \overline{X}^{(n)})}{\sum_{k=1}^n (X_{k-1}^{(n)} - \overline{X}_*^{(n)})^2}, \qquad \widetilde{\mu}_{\varepsilon,n} = \overline{X}^{(n)} - \widetilde{\alpha}_n \overline{X}_*^{(n)},$$

where

$$\overline{X}^{(n)} := \frac{1}{n}\sum_{k=1}^n X_k^{(n)}, \qquad \overline{X}_*^{(n)} := \frac{1}{n}\sum_{k=1}^n X_{k-1}^{(n)}.$$

In Section 3 we show that $(\widetilde{\alpha}_n, \widetilde{\mu}_{\varepsilon,n})_{n\in\mathbb{N}}$ is asymptotically normal, namely,

$$\begin{pmatrix} n^{3/2}(\widetilde{\alpha}_n - \alpha_n) \\ n^{1/2}(\widetilde{\mu}_{\varepsilon,n} - \mu_\varepsilon) \end{pmatrix} \xrightarrow{\mathcal{D}} \mathcal{N}(0, \Sigma_{\gamma,\varepsilon}), \qquad (2)$$

and the covariance matrix $\Sigma_{\gamma,\varepsilon}$ will be given explicitly.

It is easy to observe that the INAR(1) process is a special case of the Galton–Watson branching process with immigration if the offspring distribution is a Bernoulli distribution (see, e.g., Franke and Seligmann [9]). We recall that a Galton–Watson process is said to be subcritical, critical or supercritical if the expectation of the offspring distribution is less than 1, equals 1 or greater than 1, respectively. The result (2) can be reformulated as follows.

Corollary 1.2 *Consider a sequence of Galton–Watson branching processes with Bernoulli offspring distribution with parameter* $\alpha_n = 1 - \gamma_n/n$, $\gamma_n \to \gamma$ *where* $\gamma \geqslant 0$, *and (unobservable) immigration with expectation* $\mu_\varepsilon > 0$ *and variance* $\sigma_\varepsilon^2 < \infty$. *Then the joint conditional least squares estimator of* α_n *and* μ_ε *is asymptotically normal.*

Remark that the asymptotic normality in the sub–critical case with general offspring distribution and observed immigration is proved by Venkataraman and Nanthi [16]. The rate of convergence is $n^{1/2}$ in this case. We conjecture that our result can be extended for Galton–Watson processes with a more general offspring distribution. For this, the limit theorem (3) in Section 2 has to be generalized.

We note that Sriram [13] considered a nearly critical sequence of Galton–Watson branching processes with a general offspring distribution. However, the immigration was supposed to be observable. That is the reason why Sriram [13] investigated the limiting behaviour of another joint estimator for the offspring mean and for the mean of the immigration distribution.

2   Preliminaries

We shall need a simple lemma, which gives a sufficient condition for convergence to a functional of a continuous process. The proof is based on the Continuous Mapping Theorem (see Billingsley [4, Theorem 5.5]), and it can be found in Arató, Pap and Zuijlen [3].

The appropriate function spaces are the following Skorokhod spaces. Define $\mathbb{D}(\mathbb{R}_+, \mathbb{R}^k)$ to be the set of all functions $f : \mathbb{R}_+ \to \mathbb{R}^k$ for which $\lim_{s \uparrow t} f(s)$ exists and $f(t) = \lim_{s \downarrow t} f(s)$. The set $\mathbb{D}(\mathbb{R}_+, \mathbb{R}^k)$ can be endowed with a metric making it a complete and separable space. For measurable mappings $\Phi, \Phi_n : \mathbb{D}(\mathbb{R}_+, \mathbb{R}^k) \to \mathbb{D}(\mathbb{R}_+, \mathbb{R}^\ell)$, $n = 1, 2, \ldots$ we shall write $\Phi_n \rightsquigarrow \Phi$ if $\|\Phi_n(x_n) - \Phi(x)\|_\infty \to 0$ for all $x, x_n \in \mathbb{D}(\mathbb{R}_+, \mathbb{R}^k)$ with $\|x_n - x\|_\infty \to 0$, where $\|\cdot\|_\infty$ denotes the supremum norm.

Lemma 2.1 *Let* $\Phi, \Phi_n : \mathbb{D}(\mathbb{R}_+, \mathbb{R}^k) \to \mathbb{D}(\mathbb{R}_+, \mathbb{R}^\ell)$, $n = 1, 2, \ldots$ *be measurable mappings such that* $\Phi_n \rightsquigarrow \Phi$. *Let* Z, Z_n, $n = 1, 2, \ldots$ *be stochastic processes with values in* $\mathbb{D}(\mathbb{R}_+, \mathbb{R}^k)$ *such that* $Z_n \xrightarrow{D} Z$ *in* $\mathbb{D}(\mathbb{R}_+, \mathbb{R}^k)$ *and almost all trajectories of* Z *are continuous. Then,* $\Phi_n(Z_n) \xrightarrow{D} \Phi(Z)$ *in* $\mathbb{D}(\mathbb{R}_+, \mathbb{R}^\ell)$.

Let
$$M_k^{(n)} := X_k^{(n)} - \alpha_n X_{k-1}^{(n)} - \mu_\varepsilon.$$

Let us introduce the random step functions

$$X^{(n)}(t) := X^{(n)}_{[nt]}, \qquad M^{(n)}(t) := \sum_{k=1}^{[nt]} M_k^{(n)}, \qquad t \geqslant 0.$$

In [10] we have shown that

$$\left(\widetilde{M}^{(n)}, \widetilde{X}^{(n)} \right) := \left(\frac{M^{(n)}}{\sqrt{n}}, \frac{X^{(n)} - \mathsf{E} X^{(n)}}{\sqrt{n}} \right) \xrightarrow{\mathcal{D}} (M, X) \tag{3}$$

in the Skorokhod space $\mathbb{D}(\mathbb{R}_+, \mathbb{R}^2)$, where $(M(t))_{t \geqslant 0}$ is a time–changed Wiener process, namely, $M(t) = W(T_M(t))$ with

$$T_M(t) := \int_0^t \varrho_{\gamma, \varepsilon}(u) \, du, \qquad \varrho_{\gamma, \varepsilon}(u) := \sigma_\varepsilon^2 + \mu_\varepsilon (1 - e^{-\gamma u}),$$

and $(W(t))_{t \geqslant 0}$ is a standard Wiener process, and

$$X(t) := \int_0^t e^{-\gamma(t-s)} \, dM(s), \qquad t \geqslant 0$$

is a continuous zero mean Gaussian martingale (which is an Ornstein–Uhlenbeck type process driven by M). The main idea was first to prove that $\widetilde{M}^{(n)} \xrightarrow{\mathcal{D}} M$ by the help of the Martingale Central Limit Theorem, and then to show that $\widetilde{X}^{(n)}$ is a measurable function of $\widetilde{M}^{(n)}$, namely, $\left(\widetilde{M}^{(n)}, \widetilde{X}^{(n)} \right) = \Phi_n \left(\widetilde{M}^{(n)} \right)$ with $\Phi_n : \mathbb{D}(\mathbb{R}_+, \mathbb{R}) \to \mathbb{D}(\mathbb{R}_+, \mathbb{R}^2)$,

$$\Phi_n(x)(t) = \left(x(t), \, x\left(\frac{[nt]}{n} \right) - \gamma_n^* \int_0^{[nt]/n} e^{-\gamma_n^*([nt]/n - s)} x(s) \, ds \right),$$

where $\gamma_n^* := -n \log \alpha_n \to \gamma$. Clearly $\Phi_n \rightsquigarrow \Phi$, where

$$\Phi(x)(t) = \left(x(t), \, x(t) - \gamma \int_0^t e^{-\gamma(t-s)} x(s) \, ds \right).$$

By Lemma 2.1, $\left(\widetilde{M}^{(n)}, \widetilde{X}^{(n)} \right) \xrightarrow{\mathcal{D}} (M, X)$, since Itô's formula yields

$$\int_0^t e^{-\gamma(t-s)} \, dM(s) = M(t) - \gamma \int_0^t e^{-\gamma(t-s)} M(s) \, ds,$$

hence $(M, X) = \Phi(X)$.

Moreover, based on (3), we proved in [10] that

$$n^{3/2}(\widehat{\alpha}_n - \alpha_n) \xrightarrow{\mathcal{D}} \frac{\int_0^1 \mu_X(t) \, dM(t)}{\int_0^1 \mu_X(t)^2 \, dt} \stackrel{\mathcal{D}}{=} \mathcal{N}(0, \sigma_{\gamma,\varepsilon}^2),$$

where

$$\mu_X(t) := \mu_\varepsilon \int_0^t e^{-\gamma u} du = \begin{cases} \frac{\mu_\varepsilon}{\gamma}(1 - e^{-\gamma t}), & \gamma > 0, \\ \mu_\varepsilon t, & \gamma = 0, \end{cases}$$

$$\sigma_{\gamma,\varepsilon}^2 := \frac{\int_0^1 \mu_X(t)^2 \varrho_{\gamma,\varepsilon}(t)\, dt}{\left(\int_0^1 \mu_X(t)^2\, dt\right)^2}.$$

Introducing

$$\mu_X^{(n)}(t) := \frac{1}{n} \mathsf{E} X^{(n)}(t) = \frac{1}{n} \mathsf{E} X_{[nt]}^{(n)},$$

it is easy to show (see [10]) that $\mu_X^{(n)} \to \mu_X$ locally uniformly on $\mathbb{R}_+$, hence also in $\mathbb{D}(\mathbb{R}_+, \mathbb{R})$.

If $\alpha_n \to 1$ but not in the specified rate of n^{-1} then we have the following conjecture: assuming that the distribution of ε_1 belongs to the domain of normal attraction of a p-stable law then a similar result is valid with rate $n^{-2/p}$ and with a stable process instead of the Wiener process.

We note that Sriram [13] proved a limit theorem for the process $n^{-1}X^{(n)}$ for a nearly critical sequence of Galton–Watson branching processes with a general offspring distribution. However, the result of Sriram [13] is not applicable for a nearly critical sequence of branching processes with Bernoulli offspring distribution, since the variance $\alpha(1 - \alpha)$ of the Bernoulli distribution tends to 0 as α tends to its critical value 1. In fact, (3) implies that in this case we have $n^{-1}X^{(n)} \xrightarrow{\mathcal{D}} \mu_X$ in the Skorokhod space $\mathbb{D}(\mathbb{R}_+, \mathbb{R})$, but this limiting relationship is not sufficient for deriving the limiting behaviour of the sequence $(\tilde{\alpha}_n, \tilde{\mu}_{\varepsilon,n})$.

3   Joint Estimator

The main result of the paper is that the joint conditional least squares estimator $(\tilde{\alpha}_n, \tilde{\mu}_{\varepsilon,n})$ of the vector $(\alpha_n, \mu_\varepsilon)$ for a nearly unstable sequence of INAR(1) models is asymptotically normal.

Theorem 3.1 *Consider a sequence of INAR(1) models with parameters $\alpha_n = 1 - \gamma_n/n$ such that $\gamma_n \to \gamma$ with $\gamma \geqslant 0$, and suppose that $\mu_\varepsilon > 0$ and $\sigma_\varepsilon^2 < \infty$. Then*

$$\begin{pmatrix} n^{3/2}(\tilde{\alpha}_n - \alpha_n) \\ n^{1/2}(\tilde{\mu}_{\varepsilon,n} - \mu_\varepsilon) \end{pmatrix} \xrightarrow{\mathcal{D}} \begin{pmatrix} \dfrac{\int_0^1 \mu_X(t)\, dM(t) - \overline{\mu}_{X,1} M(1)}{\overline{\mu}_{X,2} - (\overline{\mu}_{X,1})^2} \\[3mm] \dfrac{\overline{\mu}_{X,2} M(1) - \overline{\mu}_{X,1} \int_0^1 \mu_X(t)\, dM(t)}{\overline{\mu}_{X,2} - (\overline{\mu}_{X,1})^2} \end{pmatrix} \stackrel{\mathcal{D}}{=} \mathcal{N}(0, \Sigma_{\gamma,\varepsilon}),$$

where $\overline{\mu}_{X,1} := \int_0^1 \mu_X(t)\,dt,\ \ \overline{\mu}_{X,2} := \int_0^1 (\mu_X(t))^2\,dt,\ \ and$

$$\Sigma_{\gamma,\varepsilon} = \left(\frac{\sigma_{\gamma,\varepsilon}^{(i,j)}}{\left(\overline{\mu}_{X,2} - (\overline{\mu}_{X,1})^2\right)^2} \right)_{1 \leqslant i,j \leqslant 2}$$

with

$$\sigma_{\gamma,\varepsilon}^{(1,1)} = \int_0^1 (\mu_X(t) - \overline{\mu}_{X,1})^2 \varrho_{\gamma,\varepsilon}(t)\,dt,$$

$$\sigma_{\gamma,\varepsilon}^{(1,2)} = -\int_0^1 (\mu_X(t) - \overline{\mu}_{X,1})(\overline{\mu}_{X,1}\mu_X(t) - \overline{\mu}_{X,2})\varrho_{\gamma,\varepsilon}(t)\,dt,$$

$$\sigma_{\gamma,\varepsilon}^{(2,2)} = \int_0^1 (\overline{\mu}_{X,1}\mu_X(t) - \overline{\mu}_{X,2})^2 \varrho_{\gamma,\varepsilon}(t)\,dt.$$

Proof. We have

$$\widetilde{\alpha}_n = \frac{\sum_{k=1}^n X_{k-1}^{(n)} X_k^{(n)} - n^{-1} \sum_{k=1}^n X_{k-1}^{(n)} \sum_{k=1}^n X_k^{(n)}}{\sum_{k=1}^n \left(X_{k-1}^{(n)}\right)^2 - n^{-1}\left(\sum_{k=1}^n X_{k-1}^{(n)}\right)^2},$$

hence $X_k^{(n)} - \alpha_n X_{k-1}^{(n)} = M_k^{(n)} + \mu_\varepsilon$ implies

$$\widetilde{\alpha}_n - \alpha_n = \frac{\sum_{k=1}^n X_{k-1}^{(n)} M_k^{(n)} - n^{-1} \sum_{k=1}^n X_{k-1}^{(n)} \sum_{k=1}^n M_k^{(n)}}{\sum_{k=1}^n \left(X_{k-1}^{(n)}\right)^2 - n^{-1}\left(\sum_{k=1}^n X_{k-1}^{(n)}\right)^2} = \frac{U_n}{V_n},$$

where

$$U_n := \int_0^1 X^{(n)}(t)\,dM^{(n)}(t) - M^{(n)}(1)\int_0^1 X^{(n)}(t)\,dt,$$

$$V_n := n\int_0^1 \left(X^{(n)}(t)\right)^2 dt - n\left(\int_0^1 X^{(n)}(t)\,dt\right)^2.$$

Applying $X^{(n)}(t) = n\mu_X^{(n)}(t) + n^{1/2}\widetilde{X}^{(n)}(t)$ and $M^{(n)}(t) = n^{1/2}\widetilde{M}^{(n)}(t)$, we obtain

$$U_n = U_{n,1}n^{3/2} + U_{n,2}n,$$

$$V_n = V_{n,1}n^3 + V_{n,2}n^{5/2} + V_{n,3}n^2,$$

where

$$U_{n,1} := \int_0^1 \mu_X^{(n)}(t)\,d\widetilde{M}^{(n)}(t) - \widetilde{M}^{(n)}(1)\int_0^1 \mu_X^{(n)}(t)\,dt,$$

$$U_{n,2} := \int_0^1 \widetilde{X}^{(n)}(t)\,d\widetilde{M}^{(n)}(t) - \widetilde{M}^{(n)}(1)\int_0^1 \widetilde{X}^{(n)}(t)\,dt,$$

$$V_{n,1} := \int_0^1 \left(\mu_X^{(n)}(t)\right)^2 dt - \left(\int_0^1 \mu_X^{(n)}(t)\,dt\right)^2,$$

$$V_{n,2} := 2\int_0^1 \mu_X^{(n)}(t)\widetilde{X}^{(n)}(t)\,dt - 2\int_0^1 \mu_X^{(n)}(t)\,dt\int_0^1 \widetilde{X}^{(n)}(t)\,dt,$$

$$V_{n,3} := \int_0^1 \left(\widetilde{X}^{(n)}(t)\right)^2 dt - \left(\int_0^1 \widetilde{X}^{(n)}(t)\,dt\right)^2.$$

Next we investigate

$$\widetilde{\mu}_{\varepsilon,n} = \overline{X}^{(n)} - \overline{X}_*^{(n)} \frac{\sum_{k=1}^n X_{k-1}^{(n)}X_k^{(n)} - n^{-1}\sum_{k=1}^n X_{k-1}^{(n)}\sum_{k=1}^n X_k^{(n)}}{\sum_{k=1}^n \left(X_{k-1}^{(n)}\right)^2 - n^{-1}\left(\sum_{k=1}^n X_{k-1}^{(n)}\right)^2}$$

$$= \frac{n^{-1}\sum_{k=1}^n X_k^{(n)}\sum_{k=1}^n \left(X_{k-1}^{(n)}\right)^2 - n^{-1}\sum_{k=1}^n X_{k-1}^{(n)}\sum_{k=1}^n X_{k-1}^{(n)}X_k^{(n)}}{\sum_{k=1}^n \left(X_{k-1}^{(n)}\right)^2 - n^{-1}\left(\sum_{k=1}^n X_{k-1}^{(n)}\right)^2}.$$

Clearly we have

$$\widetilde{\mu}_{\varepsilon,n} - \mu_\varepsilon = \frac{W_n}{V_n},$$

where

$$W_n := n^{-1}\sum_{k=1}^n \left(X_k^{(n)} - \mu_\varepsilon\right)\sum_{k=1}^n \left(X_{k-1}^{(n)}\right)^2 - n^{-1}\sum_{k=1}^n X_{k-1}^{(n)}\sum_{k=1}^n X_{k-1}^{(n)}\left(X_k^{(n)} - \mu_\varepsilon\right).$$

By $X_k^{(n)} - \mu_\varepsilon = \alpha_n X_{k-1}^{(n)} + M_k^{(n)}$, we can write W_n in the form

$$W_n = n^{-1}\sum_{k=1}^n M_k^{(n)}\sum_{k=1}^n \left(X_{k-1}^{(n)}\right)^2 - n^{-1}\sum_{k=1}^n X_{k-1}^{(n)}\sum_{k=1}^n X_{k-1}^{(n)}M_k^{(n)}$$

$$= M^{(n)}(1)\int_0^1 \left(X^{(n)}(t)\right)^2 dt - \int_0^1 X^{(n)}(t)\,dt\int_0^1 X^{(n)}(t)\,dM^{(n)}(t).$$

Applying again $X^{(n)}(t) = n\mu_X^{(n)}(t) + n^{1/2}\widetilde{X}^{(n)}(t)$ and $M^{(n)}(t) = n^{1/2}\widetilde{M}^{(n)}(t)$, we obtain that

$$W_n = W_{n,1}n^{5/2} + W_{n,2}n^2 + W_{n,3}n^{3/2},$$

where

$$W_{n,1} := \widetilde{M}^{(n)}(1) \int_0^1 \left(\mu_X^{(n)}(t)\right)^2 dt - \int_0^1 \mu_X^{(n)}(t)\, dt \int_0^1 \mu_X^{(n)}(t)\, d\widetilde{M}^{(n)}(t),$$

$$W_{n,2} := 2\widetilde{M}^{(n)}(1) \int_0^1 \mu_X^{(n)}(t)\widetilde{X}^{(n)}(t)\, dt - \int_0^1 \mu_X^{(n)}(t)\, dt \int_0^1 \widetilde{X}^{(n)}(t)\, d\widetilde{M}^{(n)}(t)$$

$$- \int_0^1 \widetilde{X}^{(n)}(t)\, dt \int_0^1 \mu_X^{(n)}(t)\, d\widetilde{M}^{(n)}(t),$$

$$W_{n,3} := \widetilde{M}^{(n)}(1) \int_0^1 \left(\widetilde{X}^{(n)}(t)\right)^2 dt - \int_0^1 \widetilde{X}^{(n)}(t)\, dt \int_0^1 \widetilde{X}^{(n)}(t)\, d\widetilde{M}^{(n)}(t).$$

We can notice that

$$Z_n := (U_{n,1}, U_{n,2}, V_{n,1}, V_{n,2}, V_{n,3}, W_{n,1}, W_{n,2}, W_{n,3})$$

can be expressed as a continuous function of the random vector

$$I_n := \left(\widetilde{M}^{(n)}(1), \int_0^1 \mu_X^{(n)}(t)\, dt, \int_0^1 \widetilde{X}^{(n)}(t)\, dt, \int_0^1 \left(\mu_X^{(n)}(t)\right)^2 dt, \right.$$

$$\left. \int_0^1 \left(\widetilde{X}^{(n)}(t)\right)^2 dt, \int_0^1 \mu_X^{(n)}(t)\widetilde{X}^{(n)}(t)\, dt, \int_0^1 \mu_X^{(n)}(t)\, d\widetilde{M}^{(n)}(t) \right)$$

and the random variable

$$\int_0^1 \widetilde{X}^{(n)}(t)\, d\widetilde{M}^{(n)}(t).$$

In [10] it is shown that there exist measurable functionals $\Phi, \Phi_n : \mathbb{D}(\mathbb{R}_+, \mathbb{R}) \to \mathbb{R}$, $n \in \mathbb{N}$, such that

$$\int_0^1 \mu_X^{(n)}(t)\, d\widetilde{M}^{(n)}(t) = \Phi_n(\widetilde{M}^{(n)}),$$

and $\Phi_n \rightsquigarrow \Phi$ in the sense that $|\Phi_n(x_n) - \Phi(x)| \to 0$ for all $x, x_n \in \mathbb{D}(\mathbb{R}_+, \mathbb{R})$ with $\|x_n - x\|_\infty \to 0$. Hence we conclude the existence of measurable functionals $\Psi, \Psi_n : \mathbb{D}(\mathbb{R}_+, \mathbb{R}^3) \to \mathbb{R}^7$, $n \in \mathbb{N}$, such that $I_n = \Psi_n(\mu_X^{(n)}, \widetilde{M}^{(n)}, \widetilde{X}^{(n)})$, and $\Psi_n \rightsquigarrow \Psi$ in the sense that $\|\Psi_n(x_n) - \Psi(x)\| \to 0$ for all $x, x_n \in \mathbb{D}(\mathbb{R}_+, \mathbb{R}^3)$ with $\|x_n - x\|_\infty \to 0$. Thus (3), $\mu_X^{(n)} \to \mu_X$ in $\mathbb{D}(\mathbb{R}_+, \mathbb{R})$, and an appropriate analogue of Lemma 2.1 imply $I_n \xrightarrow{D} I$ with

$$I := \left(M(1), \int_0^1 \mu_X(t)\, dt, \int_0^1 X(t)\, dt, \int_0^1 \left(\mu_X(t)\right)^2 dt, \right.$$

$$\left. \int_0^1 \left(X(t)\right)^2 dt, \int_0^1 \mu_X(t)X(t)\, dt, \int_0^1 \mu_X(t)\, dM(t) \right).$$

In [10] we have shown that

$$\int_0^1 \widetilde{X}^{(n)}(t)\, d\widetilde{M}^{(n)}(t) = A_n + B_n,$$

where

$$A_n := \frac{1}{2}\left(\widetilde{X}^{(n)}(1)\right)^2 + \frac{(1+\alpha_n)\gamma_n}{2}\int_0^1 \left(\widetilde{X}^{(n)}(t)\right)^2 dt,$$

$$B_n := \frac{1}{2n}\sum_{k=1}^n \left(M_k^{(n)}\right)^2 \xrightarrow{\;\mathcal{D}\;} \frac{1}{2}T_M(1).$$

Consequently, applying Slutsky's theorem and its corollary in Chow and Teicher [7, 8.1], we obtain $Z_n \xrightarrow{\;\mathcal{D}\;} Z$ with

$$Z := \left(U^{(1)}, U^{(2)}, V^{(1)}, V^{(2)}, V^{(3)}, W^{(1)}, W^{(2)}, W^{(3)}\right),$$

where

$$U^{(1)} := \int_0^1 \mu_X(t)\, dM(t) - M(1)\int_0^1 \mu_X(t)\, dt,$$

$$U^{(2)} := \int_0^1 X(t)\, dM(t) - M(1)\int_0^1 X(t)\, dt,$$

$$V^{(1)} := \int_0^1 \left(\mu_X(t)\right)^2 dt - \left(\int_0^1 \mu_X(t)\, dt\right)^2,$$

$$V^{(2)} := 2\int_0^1 \mu_X(t)X(t)\, dt - 2\int_0^1 \mu_X(t)\, dt \int_0^1 X(t)\, dt,$$

$$V^{(3)} := \int_0^1 \left(X(t)\right)^2 dt - \left(\int_0^1 X(t)\, dt\right)^2$$

$$W^{(1)} := M(1)\int_0^1 \left(\mu_X(t)\right)^2 dt - \int_0^1 \mu_X(t)\, dt \int_0^1 \mu_X(t)\, dM(t)$$

$$W^{(2)} := 2M(1)\int_0^1 \mu_X(t)X(t)\, dt - \int_0^1 \mu_X(t)\, dt \int_0^1 X(t)\, dM(t)$$

$$- \int_0^1 X(t)\, dt \int_0^1 \mu_X(t)\, dM(t),$$

$$W^{(3)} := M(1)\int_0^1 \left(X(t)\right)^2 dt - \int_0^1 X(t)\, dt \int_0^1 X(t)\, dM(t).$$

Again by Slutsky's argument we obtain

$$\left(n^{3/2}(\widetilde{\alpha}_n - \alpha_n),\, n^{1/2}(\widetilde{\mu}_{\varepsilon,n} - \mu_\varepsilon)\right) = \left(\frac{n^{-3/2}U_n}{n^{-3}V_n},\, \frac{n^{-5/2}W_n}{n^{-3}V_n}\right)$$

$$\xrightarrow{\;\mathcal{D}\;} \left(\frac{U^{(1)}}{V^{(1)}},\, \frac{W^{(1)}}{V^{(1)}}\right).$$

The covariance matrix $\Sigma_{\gamma,\varepsilon}$ of the limiting normal distribution can be calculated using $dM(t) = \sqrt{\varrho_{\gamma,\varepsilon}(t)}\, dW(t)$ (see [10]). This relationship implies

$$\begin{pmatrix} M(1) \\ \int_0^1 \mu_X(t)\, dM(t) \end{pmatrix} \overset{\mathcal{D}}{=} \mathcal{N}(0, \Sigma)$$

with

$$\Sigma := \begin{pmatrix} \int_0^1 \varrho_{\gamma,\varepsilon}(t)\, dt & -\int_0^1 \mu_X(t)\varrho_{\gamma,\varepsilon}(t)\, dt \\ -\int_0^1 \mu_X(t)\varrho_{\gamma,\varepsilon}(t)\, dt & \int_0^1 \left(\mu_X(t)\right)^2 \varrho_{\gamma,\varepsilon}(t)\, dt \end{pmatrix}.$$

Now the formula for $\Sigma_{\gamma,\varepsilon}$ follows, since $U^{(1)}$ and $W^{(1)}$ are linear combinations of $M(1)$ and $\int_0^1 \mu_X(t)\, dM(t)$. □

References

1. Al–Osh, M.A. and Alzaid, A.A. (1987). First–order integer–valued autoregressive (INAR(1)) process. J. Time Ser. Anal. **8**, 261–275.
2. Al–Osh, M.A. and Alzaid, A.A. (1990). An integer–valued pth–order autoregressive structure (INAR(p)) process. J. Appl. Prob. **27**, 314–324.
3. Arató, M., Pap, G. and Zuijlen, M.v. (2001). Asymptotic inference for spatial autoregression and orthogonality of Ornstein–Uhlenbeck sheets. *Computers Math. Appl.* **42**, 219–229.
4. Billingsley, P. (1968). *Convergence of Probability Measures.* John Wiley & Sons, New York.
5. Cardinal, M., Roy, R. and Lambert, J. (1999). On the application of integer–valued time series models for the analysis of disease incidence. Stat. Medicine **18**, 2025–2039.
6. Chan, N.H. and Wei, C.Z. (1987). Asymptotic inference for nearly nonstationary AR(1) processes. Ann. Statist. **15**, 1050–1063.
7. Chow, Y.S. and Teicher, H. (1978). *Probability Theory.* Springer–Verlag, New York, Heidelberg, Berlin.
8. Du, J.G. and Li, Y. (1991). The integer–valued autoregressive INAR(p) model. J. Time Ser. Anal. **12**, 129–142.
9. Franke, J. and Seligmann, T. (1993). Conditional maximum likelihood estimates for INAR(1) processes and their application to modeling epileptic seizure counts. in Subba Rao, T. (ed.), Developments in Time Series Analysis, Chapman and Hall, London, pp. 310–330.
10. Ispány, M., Pap, G. and Zuijlen, M.v. (2001). Asymptotic inference for nearly unstable INAR(1) models. Report No. 0109, University of Nijmegen, The Netherlands. Submitted to the Journal of Applied Probability.
11. Klimko, L.A. and Nelson, P.I. (1978). On conditional least squares estimation for stochastic processes. Ann. Statist. **6**, 629–642.
12. Latour, A. (1997). The multivariate GINAR(p) process. Adv. Appl. Prob. **29**, 228–248.
13. Sriram, T.N. (1994). Invalidity of bootstrap for critical branching processes with immigration. Ann. Statist. **22**, 1013–1023.

14. Steutel, F. and van Harn, K. (1979). Discrete analogues of self–decomposability and stability. Ann. Probab. **7**, 893–99.
15. Tanaka, K. (1996). *Time Series Analysis. Nonstationary and Noninvertible Distribution Theory.* John Wiley & Sons, New York.
16. Venkataraman, K.N. and Nanthi, K. (1982). A limit theorem on subcritical Galton–Watson process with immigration. Ann. Probab. **10**, 1069–1074.

Guessing the Output of a Stationary Binary Time Series

Gusztáv Morvai*

Department of Computer Science and Information Theory, Technical University of Budapest, Hungary

Abstract. The forward prediction problem for a binary time series $\{X_n\}_{n=0}^{\infty}$ is to estimate the probability that $X_{n+1} = 1$ based on the observations X_i, $0 \leq i \leq n$ without prior knowledge of the distribution of the process $\{X_n\}$. It is known that this is not possible if one estimates at all values of n. We present a simple procedure which will attempt to make such a prediction infinitely often at carefully selected stopping times chosen by the algorithm. The growth rate of the stopping times is also studied.

1 Introduction

T. Cover in [3] asked two fundamental questions concerning estimation for stationary and ergodic binary processes. Cover's first question was as follows.

Question 1 *Is there an estimation scheme f_{n+1} for the value*
$P(X_1 = 1|X_0, X_{-1}, \ldots, X_{-n})$ *such that f_{n+1} depends solely on the observed data segment $X_0, X_{-1}, \ldots, X_{-n}$ and*

$$\lim_{n \to \infty} f_{n+1}(X_0, X_{-1}, \ldots, X_{-n}) - P(X_1 = 1|X_0, X_{-1}, \ldots, X_{-n}) = 0$$

almost surely for all stationary and ergodic binary time series $\{X_n\}$?.

This question was answered by Ornstein [7] by constructing such a scheme. (See also Bailey [2].) Ornstein's scheme is not a simple one and the proof of consistency is rather sophisticated. A much simpler scheme and proof of consistency were provided by Morvai, Yakowitz, Györfi [6]. (See also Weiss [12].)

Here is Cover's second question.

Question 2 *Is there an estimation scheme f_{n+1} for the value*
$P(X_{n+1} = 1|X_0, X_1, \ldots, X_n)$ *such that f_{n+1} depends solely on the data segment $X_0, X_1, \ldots, X_n$ and*

$$\lim_{n \to \infty} f_{n+1}(X_0, X_1, \ldots, X_n) - P(X_{n+1} = 1|X_0, X_1, \ldots, X_n) = 0$$

almost surely for all stationary and ergodic binary time series $\{X_n\}$?.

* This paper has been written by the auspices of the Hungarian National Eötvös Fund. Ez a cikk a Magyar Állami Eötvös Ösztöndíj támogatásával készült.

This question was answered by Bailey [2] in a negative way, that is, he showed that there is no such scheme. (Also see Ryabko [10], Györfi, Morvai, Yakowitz [4] and Weiss [12].) Bailey used the technique of cutting and stacking developed by Ornstein [8] (see also Shields [11]). Ryabko's construction was based on a function of an infinite state Markov-chain. This negative result can be interpreted as follows. Consider a weather forecaster whose task it is to predict the probability of the event 'there will be rain tomorrow' given the observations up to the present day. Bailey's result says that the difference between the estimate and the true conditional probability cannot eventually be small for all stationary weather processes. The difference will be big infinitely often. These results show that there is a great difference between Questions 1 and 2. Question 1 was addressed by Morvai, Yakowitz, Algoet [5] and a very simple estimation scheme was given which satisfies the statement in Question refquest1 in probability instead of almost surely. Now consider a less ambitious goal than Question 2:

Question 3 *Is there a sequence of stopping times $\{\lambda_n\}$ and an estimation scheme f_n which depends on the observed data segment $(X_0, X_1, \ldots, X_{\lambda_n})$ such that*

$$\lim_{n \to \infty} f_n(X_0, X_1, \ldots, X_{\lambda_n}) - P(X_{\lambda_n+1} = 1 | X_0, X_1, \ldots, X_{\lambda_n}) = 0$$

almost surely for all stationary binary time series $\{X_n\}$?

It turns out that the answer is affirmative and such a scheme will be exhibited below. This result can be interpreted as if the weather forecaster can refrain from predicting, that is, he may say that he does not want to predict today, but will predict at infinitely many time instances, and the difference between the prediction and the true conditional probability will vanish almost surely at the stopping times.

2   Forward Estimation for Stationary Binary Time Series

Let $\{X_n\}_{n=-\infty}^{\infty}$ denote a two-sided stationary binary time series. For $n \geq m$, it will be convenient to use the notation $X_m^n = (X_m, \ldots, X_n)$. For $k = 1, 2, \ldots$, define the sequences $\{\tau_k\}$ and $\{\lambda_k\}$ recursively. Set $\lambda_0 = 0$. Let

$$\tau_k = \min\{t > 0 : X_t^{\lambda_{k-1}+t} = X_0^{\lambda_{k-1}}\}$$

and

$$\lambda_k = \tau_k + \lambda_{k-1}.$$

(By stationarity, the string $X_0^{\lambda_{k-1}}$ must appear in the sequence X_1^{∞} almost surely.) The kth estimate of $P(X_{\lambda_k+1} = 1 | X_0^{\lambda_k})$ is denoted by P_k, and is defined as

$$P_k = \frac{1}{k-1} \sum_{j=1}^{k-1} X_{\lambda_j+1} \tag{1}$$

For an arbitrary stationary binary time series $\{Y_n\}_{n=-\infty}^0$, for $k = 1, 2, \ldots$, define the sequence $\hat{\tau}_k$ and $\hat{\lambda}_k$ recursively. Set $\hat{\lambda}_0 = 0$. Let

$$\hat{\tau}_k = \min\{t > 0 : Y_{-\hat{\lambda}_{k-1}-t}^{-t} = Y_{-\hat{\lambda}_{k-1}}^0\}$$

and let

$$\hat{\lambda}_k = \hat{\tau}_k + \hat{\lambda}_{k-1}.$$

When there is ambiguity as to which time series $\hat{\tau}_k$ and $\hat{\lambda}_k$ are to be applied, we will use the notation $\hat{\tau}_k(Y_{-\infty}^0)$ and $\hat{\lambda}_k(Y_{-\infty}^0)$.

It will be useful to define another time series $\{\tilde{X}_n\}_{n=-\infty}^0$ as

$$\tilde{X}_{-\lambda_k}^0 := X_0^{\lambda_k} \text{ for all } k \geq 1. \tag{2}$$

Since $X_{\lambda_{k+1}-\lambda_k}^{\lambda_{k+1}} = X_0^{\lambda_k}$ the above definition is correct. Notice that it is immediate that $\hat{\tau}_k(\tilde{X}_{-\infty}^0) = \tau_k$ and $\hat{\lambda}_k(\tilde{X}_{-\infty}^0) = \lambda_k$.

Lemma 1 *The two time series $\{\tilde{X}_n\}_{n=-\infty}^0$ and $\{X_n\}_{n=-\infty}^\infty$ have identical distribution, that is, for all $n \geq 0$, and $x_{-n}^0 \in \{0,1\}^{n+1}$,*

$$P(\tilde{X}_{-n}^0 = x_{-n}^0) = P(X_{-n}^0 = x_{-n}^0).$$

PROOF First we prove that

$$P(\tilde{X}_{-n}^0 = x_{-n}^0, \hat{\lambda}_k(\tilde{X}_{-\infty}^0) = n) = P(X_{-n}^0 = x_{-n}^0, \hat{\lambda}_k(X_{-\infty}^0) = n). \tag{3}$$

Indeed, by (2), $\tilde{X}_{-\hat{\lambda}_k(\tilde{X}_{-\infty}^0)}^0 = X_0^{\lambda_k}$, and it yields

$$P(\tilde{X}_{-n}^0 = x_{-n}^0, \hat{\lambda}_k(\tilde{X}_{-\infty}^0) = n) = P(X_0^n = x_{-n}^0, \lambda_k = n),$$

and by stationarity,

$$P(X_0^n = x_{-n}^0, \lambda_k = n) = P(X_{-n}^0 = x_{-n}^0, \hat{\lambda}_k(X_{-\infty}^0) = n)$$

and (3) is proved. Apply (3) in order to get

$$P(\tilde{X}^0_{-n} = x^0_{-n})$$

$$= \sum_{j=n}^{\infty} P(\tilde{X}^0_{-n} = x^0_{-n}, \hat{\lambda}_n(\tilde{X}^0_{-\infty}) = j)$$

$$= \sum_{j=n}^{\infty} \sum_{x^{-n-1}_{-j} \in \{0,1\}^{j-n}} P(\tilde{X}^0_{-j} = x^0_{-j}, \hat{\lambda}_n(\tilde{X}^0_{-\infty}) = j)$$

$$= \sum_{j=n}^{\infty} \sum_{x^{-n-1}_{-j} \in \{0,1\}^{j-n}} P(X^0_{-j} = x^0_{-j}, \hat{\lambda}_n(X^0_{-\infty}) = j)$$

$$= \sum_{j=n}^{\infty} P(X^0_{-n} = x^0_{-n}, \hat{\lambda}_n(X^0_{-\infty}) = j)$$

$$= P(X^0_{-n} = x^0_{-n})$$

and Lemma 1 is proved.

Since $\{X_n\}_{n=-\infty}^{\infty}$ is a stationary time series, by Lemma 1 so is $\{\tilde{X}_n\}_{n=-\infty}^{0}$. Since a stationary time series can always be extended to be a two-sided time series we have also defined $\{\tilde{X}_n\}_{n=-\infty}^{\infty}$. Now we prove the universal consistency of the estimator P_k.

Theorem 1 *For all stationary binary time series $\{X_n\}$ and estimator defined in (1),*

$$\lim_{k \to \infty} P_k - P(X_{\lambda_k+1} = 1|X_0^{\lambda_k}) = 0 \quad almost \ surely. \tag{4}$$

Moreover,

$$\lim_{k \to \infty} P_k = \lim_{k \to \infty} P(X_{\lambda_k+1} = 1|X_0^{\lambda_k}) = P(\tilde{X}_1 = 1|\tilde{X}^0_{-\infty}) \tag{5}$$

almost surely.

PROOF

$$P_k - P(X_{\lambda_k+1} = 1|X_0^{\lambda_k})$$

$$= \frac{1}{k-1} \sum_{j=1}^{k-1} \{X_{\lambda_j+1} - P(X_{\lambda_j+1} = 1|X_0^{\lambda_j})]\}$$

$$+ \frac{1}{k-1} \sum_{j=1}^{k-1} \{P(X_{\lambda_j+1} = 1|X_0^{\lambda_j}) - P(X_{\lambda_k+1} = 1|X_0^{\lambda_k})\}$$

$$= \frac{1}{k-1} \sum_{j=1}^{k-1} \Gamma_j + \frac{1}{k-1} \sum_{j=1}^{k-1} (\Delta_j - \Delta_k).$$

Observe that $\{\Gamma_j, \sigma(X_0^{\lambda_j+1})\}$ is a bounded martingale difference sequence for $1 \leqslant j < \infty$. To see this note that $\sigma(X_0^{\lambda_j+1})$ is monotone increasing, and Γ_j is measurable with respect to $\sigma(X_0^{\lambda_j+1})$, and $E(\Gamma_j | X_0^{\lambda_{j-1}+1}) = 0$ for $1 \leqslant j < \infty$. Now apply Azuma's exponential bound for bounded martingale differences in [1] to get that for any $\epsilon > 0$,

$$P(|\frac{1}{(k-1)} \sum_{j=1}^{k-1} \Gamma_j| > \epsilon) \leq 2 \exp(-\epsilon^2(k-1)/2).$$

After summing the right side over k, and appealing to the Borel-Cantelli lemma for a sequence of ϵ's tending to zero we get

$$\frac{1}{(k-1)} \sum_{j=1}^{k-1} \Gamma_j \to 0 \quad \text{almost surely.}$$

It remains to show

$$\frac{1}{k-1} \sum_{j=1}^{k-1} \Delta_j - \Delta_k \to 0 \quad \text{almost surely.}$$

Define

$$p_{k,n}(x_{-n}^0) = P(X_{\lambda_k+1} = 1 | X_0^{\lambda_k} = x_{-n}^0, \lambda_k = n)$$

and (applying $\hat{\lambda}_k$ to the time series $\{\tilde{X}_n\}_{n=-\infty}^0$)

$$\tilde{p}_{k,n}(x_{-n}^0) = P(\tilde{X}_1 = 1 | \tilde{X}_{-\hat{\lambda}_k}^0 = x_{-n}^0, \hat{\lambda}_k = n).$$

Now the fact that $\lambda_k = \hat{\lambda}_k$ and Lemma 1 together imply

$$p_{k,n}(x_{-n}^0) = \tilde{p}_{k,n}(x_{-n}^0). \tag{6}$$

By (2) and (6),

$$\tilde{p}_{k,\lambda_k}(X_{\lambda_k}^0) = \tilde{p}_{k,\hat{\lambda}_k}(\tilde{X}_{-\hat{\lambda}_k}^0). \tag{7}$$

Combine (6) and (7) in order to get

$$P(X_{\lambda_k+1} = 1 | X_0^{\lambda_k}) = P(\tilde{X}_1 = 1 | \tilde{X}_{-\hat{\lambda}_k}^0).$$

Notice that $\{P(\tilde{X}_1 = 1 | \tilde{X}_{-\hat{\lambda}_k}^0), \sigma(\tilde{X}_{-\hat{\lambda}_k}^0)\}$ is a bounded martingale and so it converges almost surely to $P(\tilde{X}_1 = 1 | \tilde{X}_{-\infty}^0)$, and so does $P(X_{\lambda_k+1} = 1 | X_0^{\lambda_k})$. We have proved that Δ_j converges almost surely. Now Toeplitz lemma yields that $\frac{1}{k-1} \sum_{j=1}^{k-1}(\Delta_j - \Delta_k) \to 0$ almost surely. The proof of Theorem 1 is complete.

3   The Growth Rate of the Stopping Times

The next result shows that the growth of the stopping times $\{\lambda_k\}$ is rather rapid. Let $p(x^0_{-n}) = P(X^0_{-n} = x^0_{-n})$.

Theorem 2 *Let $\{X_n\}$ be a stationary and ergodic binary time series. Suppose that $H > 0$ where*

$$H = \lim_{n \to \infty} -\frac{1}{n+1} E \log p(X_0, \ldots, X_n)$$

is the process entropy. Let $0 < \epsilon < H$ be arbitrary. Then for k large enough,

$$\lambda_k(\omega) \geq c^{c^{\cdot^{\cdot^c}}} \quad \text{almost surely,} \tag{8}$$

where the height of the tower is $k - K$, $K(\omega)$ is a finite number which depends on ω, and $c = 2^{H-\epsilon}$.

PROOF Since by (2), $\lambda_k = \hat{\lambda}_k(\check{X}^0_{-\infty})$, and by Lemma 1 the time series $\{X_n\}^\infty_{-\infty}$ and $\{\check{X}_n\}^\infty_{-\infty}$ have identical distributions, and hence the same entropy, it is enough to prove the result for $\hat{\lambda}_k(\check{X}^0_{-\infty})$. Now $\hat{\tau}_k$ and $\hat{\lambda}_k$ are evaluated on the process $\{\check{X}_n\}^0_{n=-\infty}$.

For $0 < l < \infty$ define

$$R(l) = \min\{j \geq l+1 : \check{X}^{-j}_{-l-j} = \check{X}^0_{-l}\}.$$

By Ornstein and Weiss [9],

$$\frac{1}{l+1} \log R(l) \to H \quad \text{almost surely.} \tag{9}$$

First we show that if $H > 0$ then for k large enough $\hat{\tau}_{k+1} > \hat{\lambda}_k$ almost surely. We argue by contradiction. Suppose that $\hat{\tau}_{k+1} \to \infty$ and $\hat{\tau}_{k+1} \leq \hat{\lambda}_k$ infinitely often. Then

$$\check{X}^0_{-\hat{\lambda}_k} = \check{X}^{-\hat{\tau}_{k+1}}_{-\hat{\lambda}_k - \hat{\tau}_{k+1}}$$

and $\hat{\tau}_{k+1} \leq \hat{\lambda}_k$ infinitely often. Hence

$$\check{X}^0_{-\hat{\tau}_{k+1}+1} = \check{X}^{-\hat{\tau}_{k+1}}_{-\hat{\tau}_{k+1}-\hat{\tau}_{k+1}+1}$$

infinitely often and $R(\hat{\tau}_{k+1} - 1) \leq \hat{\tau}_{k+1}$ infinitely often. Then by (9),

$$H = \lim_{k \to \infty} \frac{1}{\hat{\tau}_{k+1}} \log R(\hat{\tau}_{k+1} - 1)$$

$$\leq \lim_{k \to \infty} \frac{1}{\hat{\tau}_{k+1}} \log \hat{\tau}_{k+1}$$

$$= 0$$

provided that $\hat{\tau}_k \to \infty$. Now assume that $\eta = \sup_{0<k<\infty} \hat{\tau}_k$ is finite. Then $R(n\eta - 1) = n\eta$. Now by (9),

$$
\begin{aligned}
H &= \lim_{n\to\infty} \frac{1}{n\eta} \log R(n\eta - 1) \\
&\leq \lim_{n\to\infty} \frac{1}{n\eta} \log(n\eta) \\
&= 0.
\end{aligned}
$$

We have shown that $H > 0$ implies that for k large enough $\hat{\tau}_{k+1} > \hat{\lambda}_k$ almost surely and hence for k large enough $R(\hat{\lambda}_k) = \hat{\tau}_{k+1}$ almost surely. Hence by (9),

$$
\frac{1}{\hat{\lambda}_k + 1} \log \hat{\tau}_{k+1} \to H \text{ almost surely.}
$$

Thus for almost every $\omega \in \Omega$ there exists a positive finite integer $K(\omega)$ such that for $k \geq K(\omega)$, $\frac{1}{\hat{\lambda}_k+1} \log \hat{\tau}_{k+1} > H - \epsilon$ and

$$
\hat{\lambda}_{k+1} > \hat{\tau}_{k+1} > c^{\hat{\lambda}_k} \text{ for } k \geq K(\omega)
$$

and the proof of Theorem 2 is complete.

4   Guessing the Output at Stopping Time Instances

If the weather forecaster is pressed to say simply will it rain or not tomorrow then we need a guessing scheme, rather than a predictor. Define the guessing scheme $\{\bar{X}_{\lambda_k}\}$ for the values $\{X_{\lambda_k+1}\}$ as

$$
\bar{X}_{\lambda_k} = 1_{\{P_k \geq 0.5\}}.
$$

Let $X_{\lambda_k}^*$ denote the Bayes rule, that is,

$$
X_{\lambda_k}^* = 1_{\{P(X_{\lambda_k+1}=1 | X_0^{\lambda_k}) \geq 0.5\}}.
$$

Theorem 3 Let $\{X_n\}_{n=-\infty}^{\infty}$ be a stationary binary time series. The proposed guessing scheme $\bar{X}_{\lambda_k}$ works in the average at stopping times λ_k just as well as the Bayes rule, that is,

$$
\lim_{n\to\infty} \frac{1}{n} \sum_{k=1}^{n} 1_{\{\bar{X}_{\lambda_k}=X_{\lambda_k+1}\}} - \frac{1}{n} \sum_{k=1}^{n} 1_{\{X_{\lambda_k}^*=X_{\lambda_k+1}\}} = 0 \tag{10}
$$

almost surely. Moreover,

$$
\lim_{k\to\infty} P(\bar{X}_{\lambda_k} = X_{\lambda_k+1} | X_0^{\lambda_k}) - P(X_{\lambda_k}^* = X_{\lambda_k+1} | X_0^{\lambda_k}) = 0 \tag{11}
$$

almost surely.

PROOF

$$\sum_{k=1}^{n} 1_{\{\bar{X}_{\lambda_k}=X_{\lambda_k+1}\}} - \frac{1}{n} \sum_{k=1}^{n} 1_{\{X^*_{\lambda_k}=X_{\lambda_k+1}\}} =$$

$$\frac{1}{n} \sum_{k=1}^{n} [1_{\{\bar{X}_{\lambda_k}=X_{\lambda_k+1}\}} - P(\bar{X}_{\lambda_k} = X_{\lambda_k+1}|X_0^{\lambda_k})]$$

$$- \frac{1}{n} \sum_{k=1}^{n} [1_{\{X^*_{\lambda_k}=X_{\lambda_k+1}\}} - P(X^*_{\lambda_k} = X_{\lambda_k+1}|X_0^{\lambda_k})]$$

$$+ \frac{1}{n} \sum_{k=1}^{n} [P(\bar{X}_{\lambda_k} = X_{\lambda_k+1}|X_0^{\lambda_k}) - P(X^*_{\lambda_k} = X_{\lambda_k+1}|X_0^{\lambda_k})]$$

$$= \Gamma_n + \Theta_n + \Psi_n.$$

Now Γ_n and Θ_n tend to zero since they are averages of bounded martingale differences (cf. Azuma [1]). Concerning the third term Ψ_n, it is enough to prove that

$$\lim_{k \to \infty} P(\bar{X}_{\lambda_k} = X_{\lambda_k+1}|X_0^{\lambda_k})] - P(X^*_{\lambda_k} = X_{\lambda_k+1}|X_0^{\lambda_k}) = 0$$

almost surely. To see this recall the result in Theorem 1,

$$\lim_{k \to \infty} P_k = \lim_{k \to \infty} P(X_{\lambda_k+1} = 1|X_0^{\lambda_k}) = P(\tilde{X}_1 = 1|\tilde{X}_{-\infty}^0)$$

almost surely, and apply this in order to get

$$\lim_{k \to \infty} [P(\bar{X}_{\lambda_k} = X_{\lambda_k+1}|X_0^{\lambda_k}) - P(X^*_{\lambda_k} = X_{\lambda_k+1}|X_0^{\lambda_k})] =$$

$$\lim_{k \to \infty} [P(P(\tilde{X}_1 = 1|\tilde{X}_{-\infty}^0) \neq 0.5, \bar{X}_{\lambda_k} = X_{\lambda_k+1}|X_0^{\lambda_k})$$

$$- \quad P(P(\tilde{X}_1 = 1|\tilde{X}_{-\infty}^0) \neq 0.5, X^*_{\lambda_k} = X_{\lambda_k+1}|X_0^{\lambda_k})]$$

$$+ \quad [P(P(\tilde{X}_1 = 1|\tilde{X}_{-\infty}^0) = 0.5, \bar{X}_{\lambda_k} = X_{\lambda_k+1}|X_0^{\lambda_k})$$

$$- \quad P(P(\tilde{X}_1 = 1|\tilde{X}_{-\infty}^0) = 0.5, X^*_{\lambda_k} = X_{\lambda_k+1}|X_0^{\lambda_k})]$$

$$= \quad 0.$$

The proof of Theorem 3 is now complete.

Acknowledgements

The author wishes to thank Benjamin Weiss for helpful discussions and suggestions. This paper has been written by the auspices of the Hungarian National Eötvös Fund. (Ez a cikk a Magyar Állami Eötvös Ösztöndíj támogatásával készült.)

References

1. Azuma, K. (1967), Weighted sums of certain dependent random variables. Tohoku Math. J. **37**, 357–367
2. Bailey, D.H. (1976). Sequential Schemes for Classifying and Predicting Ergodic Processes. Ph. D. thesis, Stanford University
3. Cover, T.M. (1975). Open problems in information theory. In: 1975 IEEE Joint Workshop on Information Theory. 35–36. IEEE Press, New York
4. Györfi, L., Morvai, G., Yakowitz, S. (1998). Limits to consistent on-line forecasting for ergodic time series. IEEE T. Inf. Theory **44**, 886–892
5. Morvai, G., Yakowitz, S., Algoet, P. (1997). Weakly convergent nonparametric forecasting of stationary time series. IEEE T . Inf. Theory **43**, 483–498
6. Morvai, G., Yakowitz, S., Györfi, L. (1996). Nonparametric inferences for ergodic, stationary time series. Ann. Stat. **24**, 370–379
7. Ornstein, D.S. (1978). Guessing the next output of a stationary process. Israel J. Math. **30**, 292–296
8. Ornstein, D.S. (1974). *Ergodic Theory, Randomness, and Dynamical Systems.* Yale University Press
9. Ornstein, D.S., Weiss, B. (1993). Entropy and data compression schemes. IEEE T. Inf. Theory **39**, 78–83
10. Ryabko, B.Ya. (1988). Prediction of random sequences and universal coding. Problems of Inform. Trans. **24**, 87–96
11. Shields, P.C. (1991). Cutting and stacking: a method for constructing stationary processes. IEEE T. Inf. Theory **37**, 1605–1614
12. Weiss, B. (2000). *Single Orbit Dynamics.* CBMS 95. American Mathematical Society, Providence

Asymptotic Expansions for Long-Memory Stationary Gaussian Processes

David M. Zucker[1], Judith Rousseau[2], Anne Philippe[3] and Offer Lieberman[4]

[1] Hebrew University of Jerusalem, Israel
[2] University of Paris V, France
[3] University of Lillé I, France
[4] Technion-Israel Institute of Technology, Israel

Abstract. This paper surveys results recently obtained by the authors on higher-order asymptotic expansions for stationary Gaussian processes with long memory, that is, with a hyperbolically decaying autocovariance function. Such processes have been used to model time series data in various fields. Frequentist-type results presented include the following: an Edgeworth expansion for the sample autocovariance function, an Edgeworth expansion for the log-likelihood derivatives and the maximum likelihood estimator in parametric time series models, and a Bartlett corrected likelihood ratio test for the fractional integration parameter in the ARFIMA model. Bayesian-type results presented include the following: an Edgeworth expansion for the posterior density of the parameter vector in parametric models, identification of matching priors under which frequentist and Bayesian inferences approximately agree, and identification of approximate reference priors in the sense of Bernardo, which carry minimum initial information on the parameter vector in a certain Kullback-Leibler sense. The key tools are theorems concerning the limiting behavior of the trace of the product of certain Toeplitz matrices and a general theorem of Durbin on Edgeworth expansions for dependent data. The results and proofs are briefly sketched, with references to the original papers for further details.

1 Introduction

This paper reviews recent results obtained by the authors on asymptotic expansions for a long-memory stationary Gaussian process $\{X_t, t \in \mathbb{Z}\}$. As discussed by Brockwell and Davis ([9], Sec. 13.2), a stationary process may be classified as short, intermediate, or long memory on the basis of the behavior of the autocovariance function $\gamma(u)$ as follows (K being a constant):

Short memory: $\gamma(u) \sim K\rho^{-|u|}$ with $\rho \in (0,1)$ as $u \to \infty$
Intermediate memory: $\gamma(u) \sim K|u|^{2d-1}$ with $d \in (-\frac{1}{2}, 0)$ as $u \to \infty$
Long memory: $\gamma(u) \sim K|u|^{2d-1}$ with $d \in (0, \frac{1}{2})$ as $u \to \infty$

Under long memory, the autcovariance function is not absolutely summable. Long memory Gaussian processes have been used to model long range dependence in various fields, such as hydrology, economics, and finance (see Robinson [27]).

Very slow decay in the autocovariance function leading to nonsumma-bility corresponds to a pole at the origin in the spectral density function. Accordingly, Robinson and others have used the term strongly dependent process or long memory process to refer to a process whose spectral density $f(\lambda)$ satisfies

$$f(\lambda) \sim |\lambda|^{-\alpha} A(\lambda) \text{ as } \lambda \to 0, \tag{1}$$

with $0 < \alpha < 1$ and $A(\lambda)$ slowly varying at 0 in the sense that $\lambda^{\delta} A(\lambda)$ is bounded for every $\delta > 0$. For parametric models, α and A usually will depend on the model parameter vector θ. The quantity α in (1) corresponds to $2d$ for d as in the Brockwell-Davis formulation above.

The most well known long memory time series model is the fractionally integrated autoregressive - moving average (ARFIMA) model (Granger and Joyaeux [16]; Hosking [19]). The ARFIMA model is an extension of the clas-sical Box-Jenkins ([8]) ARMA model, which is a short memory model. With B denoting the backshift operator ($BX_t = BX_{t-1}$), the ARFIMA model is defined by $\Phi(B)(1-B)^d X_t = \Psi(B)\varepsilon_t$, where the ε_t's are i.i.d. $N(0,\sigma^2)$, Φ and Ψ are polynomials (whose coefficients are parameters of the process), and $d \in (-\frac{1}{2}, \frac{1}{2})$. When $d = 0$ we get an ARMA model. For $d \neq 0$ we have $\gamma(u) \sim K|u|^{2d-1}$ and the process is intermediate or long memory according to whether $d \in (-\frac{1}{2}, 0)$ or $d \in (0, \frac{1}{2})$.

Our work deals with the sample autocovariance function (SACF), which is a key tool in preliminary analysis of time series data, and with the maximum likelihood estimator (MLE) in parametric time series models such as the ARFIMA model. Both involve quadratic forms in correlated normal random variables. Prior literature provides asymptotic normality results for the SACF and the MLE for long memory Gaussian processes: the SACF and related quadratic forms have been treated by Fox and Taqqu ([15]) and Avram ([2]), while the MLE has been treated by Dahlhaus ([10]). We take the analysis a step further by developing Edgeworth-type asymptotic expansions.

We present both frequentist and Bayesian results. The frequentist results are in the form of asymptotic approximations to sampling distributions. The Bayesian results are in the form of asymptotic approximations to posterior distributions. In the Bayesian setting we also discuss matching priors in the sense of Welch and Peers ([31]) and Peers ([25]), *i.e.*, priors under which posterior probabilities equal frequentist p-values up to some order of approx-imation, and reference priors in the sense of Bernardo ([5]).

Our main tool is a theorem of Durbin ([12]) giving conditions for valid-ity of the Edgeworth expansion for dependent data. Some modifications to Durbin's theorem are necessary to apply the theorem in our setting. In veri-fying Durbin's conditions, we have to deal with traces of products of Toeplitz matrices. We exploit results of Fox and Taqqu ([14]) and Dahlhaus ([10]) on the behavior of such quantities. For the analysis of the MLE, we establish a uniform version of these Toeplitz matrix theorems.

2 Background

We work with n consecutive observations from the process $\{X_t\}$. The data vector is represented by $x = (X_1, \ldots, X_n)'$. The autocovariance function (ACF) $\gamma(u)$ and spectral density function $f(\lambda)$ are related by $\gamma(u) = \mathcal{F}(f)(u)$, where

$$\mathcal{F}(h)(u) = \int_{-\pi}^{\pi} h(\lambda)e^{i\lambda u}d\lambda.$$

The basic model assumption is that $x \sim N(0, \Sigma_n)$, where $\Sigma_n = T_n(f)$, with $T_n(h) = [\mathcal{F}(h)(j - k)]_{1 \leqslant j,k \leqslant n}$. In this work we deal with the mean zero case; O. Lieberman and D. Andrews are currently working on extension to the non-zero mean case. The sample autocovariance function (SACF) is

$$\hat{\gamma}(u) = \frac{1}{n-u} \sum_{t=1}^{n-u} X_t X_{t+u}$$

This function is a common tool for initial examination of data.

The log likelihood in a parametric model with parameter vector θ is

$$L(\theta) = -\frac{n}{2}\log 2\pi - \frac{1}{2}\log\det(\Sigma_n) - \frac{1}{2}x'\Sigma_n^{-1}x.$$

The analysis of the MLE of θ involves analysis of the log likelihood derivatives (LLD's).

Both the SACF and the LLD's are quadratic forms in x. The SACF for $u > 0$ can be written as $\hat{\gamma}(u) = (n - u)^{-1}x'A_{n,u}\,x$ with $[A_{n,u}]_{j,k}$ equal to $\frac{1}{2}$ when $|j - k| = u$ and equal to 0 otherwise. It may be shown that $A_{n,u} = T_n(g_u)$ with $g_u(\lambda) = \cos(u\lambda)/(2\pi)$. For $u = 0$, $\hat{\gamma}(0) = n^{-1}x'x$.

The LLD $\partial L(\theta)/\partial\theta_{r_1}\cdots\partial\theta_{r_q}$ takes the form

$$L_\nu = x'B_\nu(\theta)x - F_\nu(\theta), \qquad \nu = (r_1\ldots r_q),$$

with

$$B_\nu(\theta) = -\frac{1}{2}\frac{\partial^q \Sigma_n^{-1}}{\partial\theta_{r_1}\cdots\partial\theta_{r_q}}$$

$$F_\nu(\theta) = -\frac{1}{2}\frac{\partial^q \log\det(\Sigma_n)}{\partial\theta_{r_1}\cdots\partial\theta_{r_q}}.$$

Using classical matrix derivative results (Harville, [18]), it may be seen that

$$B_\nu(\theta) = \Sigma_{k=1}^{b_{1\nu}}a_{1k}\left[\Pi_{j=1}^{p_{1k}}T_n^{-1}(f_\theta)T_n(g_{1,\theta,j})\right]T_n^{-1}(f_\theta)$$

$$F_\nu(\theta) = \Sigma_{k=1}^{b_{2\nu}}a_{2k}\,tr\left[\Pi_{j=1}^{p_{2k}}T_n^{-1}(f_\theta)T_n(g_{2,\theta,j})\right]$$

where the $g_{m,\theta,j}$'s are derivatives of the spectral density with respect to θ.

3   Basic Frequentist Results

Let ω denote a parameter of interest and $\hat{\omega}$ a corresponding estimate. In our setting, ω may represent a vector of ACF values $\gamma(u)$ for a fixed set of u's or the parameter θ of a parametric model, with $\hat{\omega}$ being correspondingly a vector of SACF values or the MLE of θ. Classical first-order asymptotics typically leads to a result to the effect that $\sqrt{n}(\hat{\omega} - \omega)$ is approximately distributed as $N(0, M)$ for large n. An Edgeworth expansion is a result of the form

$$\Pr(\sqrt{n}(\hat{\omega} - \omega)) \in C) = \int_C \varphi_M(u)\Gamma_{n,r}(u)du + o(n^{-\frac{1}{2}s+1}) \qquad (2)$$

with C being any Borel set, φ_M being the $N(0, M)$ density, and

$$\Gamma_{n,r}(u) = 1 + \Sigma_{r=3}^s n^{-\frac{1}{2}r+1} q_{n,r}(u),$$

where $q_{n,r}$ are polynomials whose coefficients involve normalized cumulants of $\hat{\omega}$.

Edgeworth expansion is a classical topic. Some basic references are Bhattacharya and Rao ([7]), which gives a comprehensive treatment for sums of i.i.d. random variables and vectors, and Barndorff-Nielsen and Cox (see [4], Chapters 4 and 6). In particular, these references discuss the form of the polynomials $q_{n,r}$. Taniguchi ([29]) and Taniguchi and Kakizawa ([30]) discuss Edgeworth expansions for ARMA and other short memory time series models (these monographs summarize a series of papers by Taniguchi and co-workers in this area). We have obtained the following results for the long memory case, that is, for a process satisfying (1) with $0 \leqslant \alpha < 1$.

Theorem 1. *Under suitable technical conditions, the vector of SACF values $\hat{\gamma}(u)$ for a fixed finite set of u values admits a valid Edgeworth expansion of the form (2).*

Theorem 2. *Under suitable technical conditions, the vector of LLD's (omitting redundant ones) and the MLE vector admit a valid Edgeworth expansion of the form (2).*

Theorem 1 on the SACF is proven in Lieberman, Rousseau, and Zucker ([23]), which also gives a result on more general quadratic forms. A similar result holds for the sample autocorrelation function $\hat{\rho}(u) = \hat{\gamma}(u)/\hat{\gamma}(0)$. Theorem 2 on the LLD's and MLE is proven in Lieberman, Rousseau, and Zucker ([24]). The next two sections briefly describe the main pieces of the proof: a theorem on products of Toeplitz matrices and Durbin's Edgeworth expansion theorem for dependent data (see [12]). For the analysis of the MLE, an Edgeworth expansion for the LLD's is established first, and then arguments along the lines of Bhattacharya and Ghosh ([6]), based on a Taylor expansion of the first-order LLD's, are used to pass to the Edgeworth expansion results for the MLE (see [24] for details).

4   A Result on Products of Toeplitz Matrices

In Theorems 3 and 4 we present results on the limiting behavior of the trace of the product of Toeplitz matrices (TPTM). Such quantities appear extensively in the asymptotic analysis of the SACF and the MLE, as will become apparent in the next section. The behavior of such quantities is a topic of longstanding interest going back to Grenander and Szegö ([17]). Taniguchi ([28]) presented results in the ARMA setting. For the long memory setting, the non-uniform versions of Theorems 3 and 4 are due to Fox and Taqqu's Theorem 1.a in [15] and Dahlhaus's Theorem 5.1, respectively. The uniform version, which is needed in the analysis of the MLE, is proven in [24].

Theorem 3. *Let Θ be an open subset of $\mathbb{R}^m$. For $\theta \in \Theta$, let $f_{\theta,1}, ..., f_{\theta,p}$ and $g_{\theta,1}, ..., g_{\theta,p}$ be symmetric real-valued functions defined on $[-\pi, \pi]$ and continuous on $\{\lambda : |\lambda| > t\}$, $\forall t > 0$. Suppose that $\forall \theta \in \Theta, \exists \varepsilon > 0$, such that $\forall \delta > 0, \exists M_\theta \geqslant 0$ for which $\sup_{|\theta'-\theta|<\varepsilon} |f_{\theta',i}(\lambda)| \leqslant M_\theta |\lambda|^{-\alpha(\theta)-\delta}$, $i = 1, ..., p$, and $\sup_{|\theta'-\theta|<\varepsilon} |g_{\theta',i}(\lambda)| \leqslant M_\theta |\lambda|^{-\beta(\theta)-\delta}$, $i = 1, ..., p$, for all $\lambda > 0$, where $\alpha(\theta) < 1$ and $\beta(\theta) < 1$. Also suppose that $\forall t > 0, \exists M_{t,\theta}$ such that*

$$\sup_{|\theta'-\theta|<\varepsilon; |\lambda|>t} \left| \frac{df_{\theta',i}(\lambda)}{d\lambda} \right| \leqslant M_{t,\theta} \quad and \quad \sup_{|\theta'-\theta|<\varepsilon; |\lambda|>t} \left| \frac{dg_{\theta',i}(\lambda)}{d\lambda} \right| \leqslant M_{t,\theta},$$

$$i = 1, ..., p.$$

Assume further that $p(\alpha(\theta)+\beta(\theta)) < 1$ for all θ. Then for any compact subset Θ^* of Θ,

$$\lim_{n\to\infty} \sup_{\theta'\in\Theta^*} \left| \frac{1}{n} tr \left[\prod_{i=1}^{p} T_n(f_{\theta',i}) T_n(g_{\theta',i}) \right] - (2\pi)^{2p-1} \int_{-\pi}^{\pi} \prod_{i=1}^{p} f_{\theta',i}(\lambda) g_{\theta',i}(\lambda) d\lambda \right|$$

$$= 0.$$

Theorem 4. *Let Θ be a open subset of $\mathbb{R}^m$ and let Θ^* be a compact subset of Θ. Let p be a positive integer and let $\alpha(\theta)$ and $\beta(\theta)$ be continuous functions on Θ^* with range in (0,1) satisfying $\beta(\theta) - \alpha(\theta) < 1/(2p)$. Suppose that $f_{\theta,j}(\lambda), j \leqslant p$, are symmetric nonnegative functions and that $g_{\theta,j}(\lambda), j \leqslant p$, are symmetric real-valued functions satisfying the following conditions:*

A. Each $f_{\theta,j}(\lambda)$ is differentiable with respect to θ, with the derivative being continuous in (λ, θ) over $\theta \in \Theta$ and $\lambda \neq 0$. The function $f_{\theta,j}(\lambda)^{-1}$ is continuous in (λ, θ) for all λ and θ. The derivatives $(\partial/\partial\lambda) f_{\theta,j}(\lambda)^{-1}$ and $(\partial^2/\partial\lambda^2) f_{\theta,j}(\lambda)^{-1}$ are continuous in (λ, θ) for $\lambda \neq 0$. There exist $\alpha(\theta) \in (0, 1)$ and $c_1(\delta), c_2(\delta)$ such that for all $\delta > 0$

$$\left| \left(\frac{\partial}{\partial\lambda} \right)^k f_{\theta,j}(\lambda)^{-1} \right| \leqslant c_1(\theta, \delta) |\lambda|^{\alpha(\theta)-k-\delta}, \quad k = 0, 1, 2,$$

and $|f_{\theta,j}(\lambda)| \leqslant c_2(\delta)|\lambda|^{-\alpha(\theta)-\delta}$ over $\lambda \in (0,\pi)$.

B. The $g_{\theta,j}$'s are continuous at all $\lambda \neq 0$ and for each $\delta > 0$ there exists $c^*(\delta)$ such that $|g_{\theta,j}(\lambda)| \leqslant c^*(\delta)|\lambda|^{-\beta(\theta)-\delta}$ for all $\theta \in \Theta^*$.

Then

$$\lim_{n\to\infty} \sup_{\theta\in\Theta^*} \left| \frac{1}{n} tr \left[\prod_{j=1}^{p} \{T_n(f_{\theta,j})^{-1}T_n(g_{\theta,j})\} \right] - \frac{1}{2\pi} \int_{-\pi}^{\pi} \prod_{j=1}^{p} (f_{\theta,j}(\lambda))^{-1} g_{\theta,j}(\lambda) d\lambda \right|$$
$$= 0.$$

In our application of these theorems, the $f_{\theta,j}$ will be the spectral density function of the process. In the SACF case, the functions $g_{\theta,j}$ will be the function $g_u(\lambda) = \cos(u\lambda)/(2\pi)$. In the MLE case, the function $f_{\theta,j}$ will be derivatives of the spectral density function. In the MLE case, part of the analysis involves the spectral density function and its derivatives for values of θ different from, but in a neighbourhood of, the true value.

5   Application of Durbin's Theorem

Durbin in [12] presents a general theorem on the validity of Edgeworth expansions for dependent data. Let S_n be a random vector with cumulants of order $O(n)$. Put $W_n = n^{-\frac{1}{2}}(S_n - E[S_n])$ and $D_n = \text{Cov}(W_n)$. Also let $\phi_n(z)$ denote the characteristic function of S_n. Durbin shows that W_n admits a valid Edgeworth expansion if D_n converges to a positive definite matrix and if certain conditions on the behavior of $\phi_n(z/\sqrt{n})$ and derivatives of $\log \phi_n(z/\sqrt{n})$ are satisfied. Durbin's proof is very similar to that in Feller ([13]) for the case of sums of i.i.d. random variables.

For the SACF and LLD's, Sec. 2 shows that we need to work with $S_{nj} = x'C_j x - \mu_j$, where C_j is a product of Toeplitz matrices. By classical multivariate normal theory (see Anderson [1]) we find that

$$D_n(j,k) = \frac{2}{n} tr(\tilde{C}_j \tilde{C}_k), \quad \tilde{C}_j = C_j \Sigma_n,$$

$$\phi_n(z) = \exp(i \Sigma_{j=1}^{m} \mu_j z_j) \det \left[I_n - 2i \Sigma_{j=1}^{m} z_j \tilde{C}_j \right]^{-\frac{1}{2}}$$

It may be seen further that the cumulants of S_n are of TPTM form, and thus of order $O(n)$ by the theorems in the preceding section. We now discuss briefly the verification of Durbin's Assumptions in this context. See [23] and [24] for the full details.

Durbin's Assumption 1 is that D_n converges to a positive definite matrix. Because the elements of D_n are TPTM's, the theorems in the preceding section imply convergence to some matrix D. For the SACF, positive definiteness is shown by a simple argument. For the MLE, [24] gives a condition for positive definiteness that is easily checked.

Durbin's Assumption 2 is as follows: For any $\delta > 0$, we have, uniformly over θ,

$$\int_{\{||z|| > \delta \sqrt{n}\}} |\phi_n(z/\sqrt{n}, \theta)| dz = o(n^{-(s-2)/2}).$$

The idea of the proof is as follows. Let $\rho_1, \ldots, \rho_n$ be the eigenvalues of $\Sigma_{j=1}^m z_j \tilde{C}_j$. Then $|\phi_n(z/\sqrt{n}, \theta)|^{-4}$ equals

$$1 + \frac{4}{n} \sum_r \rho_r^2 + \left(\frac{4}{n}\right)^2 \sum_{r \neq r'} \rho_r^2 \rho_{r'}^2 + \cdots + \left(\frac{4}{n}\right)^n \prod_{r=1}^n \rho_r^2 \qquad (3)$$

We have

$$\frac{4}{n} \sum_r \rho_r^2 = \frac{4}{n} tr \left(\sum_{j=1}^m z_j \tilde{C}_j\right)^2 = 2 z' D_n(\theta) z.$$

This quantity converges to $2z'D(\theta)z$. The remaining terms in (3) may be bounded by quantities involving TPTM's and shown to be negligible. This yields an adequate lower bound for $|\phi_n(z/\sqrt{n}, \theta)|$.

For analysis of the LLD's, for certain technical reasons we work with a modified version of Durbin's Assumption 3, as follows:

a. For any $r = (r_1, \ldots, r_d)$, with $|r| \leqslant s$, the derivatives $\partial^{|r|} \log \phi_n(\omega; \theta)/\partial \omega^r$ exist for ω in a neighbourhood of the origin.

b. For r as above, the quantity $n^{-1} \partial^{|r|} \log \phi_n(0; \theta)/\partial \omega^r$ has a limit as $n \to \infty$.

c. For any vector ξ with $||\xi|| = 1$, the quantity $n^{-1} d^s \phi_n(y\xi; \theta)/dy^s$ has a limit as $n \to \infty$ and $y \to 0$, with convergence uniform over ξ.

In [24] it is shown, by tracing step by step through Durbin's proof, that Durbin's result goes through with this assumption replacing his original Assumption 3. For analysis of the SACF, we need to work with a more elaborate modified version of Durbin's Assumption 3; see [23] for details.

The main job is to prove (c). The idea of the proof is as follows. We have

$$\log \phi_n(y\xi) = iy \sum_{j=1}^m \xi_j \mu_j - \frac{1}{2} \log \det(\Omega(y))$$

with $\Omega(y) = I - yG$, where $G = 2i \Sigma_{j=1}^m \xi_j \tilde{C}_j$. Thus

$$\frac{d}{dy} \log \phi_n(y\xi) = i \Sigma_{j=1}^m \xi_j \mu_j + \frac{1}{2} tr(\Omega^{-1} G)$$

In examining the higher order derivatives of $\log \phi_n(y\xi)$, the main type of term to deal with looks like

$$n^{-1} tr(\Omega^{-1} G \ldots \Omega^{-1} G)$$

Such terms may handled using the TPTM theorem.

6 Bartlett-Corrected Likelihood Ratio Tests

Lieberman, Rousseau, and Zucker in [22] discuss small-sample inference for
the the fractional integration parameter d in the ARFIMA(p, d, q) model.
In particular, they present a Bartlett-corrected likelihood ratio (LR) test
for the hypothesis $H_0 : d = d_0$. Bartlett correction involves expressing the
expectation of the LR statistic $\Lambda(d)$ in an expansion $E[\Lambda(d)] = 1 + H(\theta) + O(n^{-\frac{3}{2}})$, where $H(\theta)$ is $O(1/n)$, and then forming a corrected LR statistic
$\tilde{\Lambda}(d) = \{1 + H(\theta)\}^{-1}\Lambda(d)$. Here θ denotes the vector comprising all the
ARFIMA model parameters. In practice, one uses an estimate of $H(\theta)$.

Barndorff-Nielsen and Cox in [4] showed in a general likelihood setting
that if the LLD's and MLE admit an Edgeworth expansion, then the corrected statistic is distributed as χ_1^2 under H_0 to order $O(n^{-2})$ (as opposed
to $O(n^{-1})$ for the uncorrected statistic). The Edgeworth expansion results of
[24] establish the necessary conditions for this result to apply.

Lawley ([21]) and DiCiccio and Stern ([11], Eqn. (14)) presented a general
formula for $H(\theta)$ in the general maximum likelihood setting. For testing d in
the ARFIMA model, [22] provide expressions for the expected log-likelihood
derivatives appearing in this formula and for the derivatives of the autocovariance function that appear in these expected log-likelihood derivatives.
They present a numerical study showing that the accuracy of the LR test is
improved substantially with the correction.

7 Bayesian Asymptotics for Parametric Models

We now discuss our Bayesian results for parametric models. Let $\pi(\theta)$ be a
prior density for θ (positive at the true value θ_0). The likelihood function is

$$p_\theta^n(x) = \frac{e^{-x'\Sigma_n^{-1}x/2}}{\det[\Sigma_n]^{1/2}(2\pi)^{n/2}}$$

and the posterior density of θ is

$$\Pi(\theta|x) = \int \pi(\theta)p_\theta^n(x)d\theta.$$

The main result is as follows.

Theorem 5. *Under the same type of conditions as in Theorem 2, for any
Borel set C we have, up to an error of $o_{P_{\theta_0}}(n^{-\frac{1}{2}s+1})$,*

$$\int_C \Pi(\theta|x)d\theta \doteq \int_C \varphi_{\tilde{M}}\left(\frac{u - \hat{\theta}}{\sqrt{n}}\right)\Gamma_{n,r}^*\left(\frac{u - \hat{\theta}}{\sqrt{n}}\right)du,$$

*where $\hat{\theta}$ is the MLE of θ, $\tilde{M}$ is the inverse of n^{-1} times the observed infor-
mation matrix, and*

$$\Gamma_{n,r}^*(u) = 1 + \Sigma_{r=3}^s n^{-\frac{1}{2}r+1}q_{n,r}^*(u),$$

where $q^*_{n,r}$ are polynomials whose coefficients involve x.

This result is proved in [26] (PR). The proof follows the pattern of Johnson ([20]) for the i.i.d. case, and is based on Laplace expansion of the posterior integral. A Taylor expansion of the log-likelihood function is used. Terms in the Taylor expansion involve quantities of TPTM form. Use is made of frequentist results at certain stages of the proof.

Using frequentist and Bayesian asymptotic results, we can express frequentist probabilities as a function of posterior probabilities in an approximate sense. In particular, for a given parameter of interest, under a special "matching prior" we have p-value = posterior probability + $O(n^{-1})$. The matching prior for parameter j is the solution to the equation

$$\sum_k \frac{M^{jk}(\theta)}{\sqrt{M^{jj}(\theta)}} \frac{\partial \log \pi(\theta)}{\partial \theta_k} = \sum_k \frac{\partial}{\partial \theta_k} \left(\frac{M^{jk}(\theta)}{\sqrt{M^{jj}(\theta)}} \right)$$

where M^{jk} is the j, k-th element of the inverse of the limiting expected information matrix, *i.e.*, the limit as n goes to infinity of the expected information matrix of n observations divided by n. For the one-dimensional case, we get the Jeffreys prior. In higher dimensions, the situation is generally very complicated. For the ARFIMA$(0, d, 0)$ model, however, we get the following closed form solution:

$$\pi(d, \sigma) = \sigma e^{-cd} h(\sigma^2 e^{-cd}),$$

where c is a known constant and h is any smooth function. See [26] for further discussion.

Philippe and Rousseau have also obtained an asymptotic expansion for the Kullback-Leibler divergence between the density $p^n_\theta(x)$ (as a function of x) and the marginal distribution of x (after integrating over the prior). This allows for the construction of a reference prior in the sense of Bernardo ([5]), *i.e.*, a prior that maximizes the asymptotic Kullback-Leibler divergence. See [26] for details.

References

1. Anderson, T.W. (1984). *An Introduction to Multivariate Statistical Analysis*, 2nd ed. Wiley, New York.
2. Avram, F. (1988). On bilinear forms in Gaussian random variables and Toeplitz matrices. Probab. Theory Rel. **79**, 37–45
3. Barndorff-Nielsen, O.E., Hall, P. (1988). On the level-error after Bartlett adjustment of the likelihood ratio statistic. Biometrika **75**, 374–378
4. Barndorff-Nielsen, O.E., Cox, D.R. (1989). *Asymptotic Expansions for Use in Statistics*. Chapman and Hall, London
5. Bernardo, J. (1979). Reference prior distributions for Bayesian inference (with discussion). J. Roy. Stat. Soc. B **41**, 113–147

6. Bhattacharya, R.N., Ghosh, J.K (1978). On the validity of the formal Edgeworth expansion. Ann. Stat. **6**, 434–451

7. Bhattacharya, R.N., Rao, R.R. (1976). *Normal Approximation and Asymptotic Expansions*. Wiley, New York

8. Box, G.E., Jenkins, G.M. (1976). *Time Series Analysis: Forecasting and Control.* Holden-Day, San Francisco

9. Brockwell, P.J., Davis, R.A. (1991). *Time Series: Theory and Methods.* 2nd ed. Springer, New York

10. Dahlhaus, R. (1989). Efficient parameter estimation for self-similar processes. Ann. Stat. **17**, 1749–1766

11. DiCiccio, T.J., Stern, S.E. (1994). Frequentist and Bayesian Bartlett correction of test statistics based on adjusted profile likelihoods. J. Roy. Stat. Soc. B **56**, 397–408

12. Durbin, J. (1980). Approximations for densities of sufficient statistics. Biometrika **67**, 311–333

13. Feller, W. (1971). *An Introduction to Probability Theory and its Applications* **2**, 2nd ed. Wiley, New York

14. Fox, R., Taqqu, M.S. (1986). Large sample properties of parameter estimates for strongly dependent stationary Gaussian time series. Ann. Stat. **14**, 517–532

15. Fox, R., Taqqu, M.S. (1987). Central limit theorems for quadratic forms in random variables having long-range dependence. Probab. Theory Rel. **74**, 213–240

16. Granger, C.W., Joyeux, R. (1980). An introduction to long-memory time series and fractional differencing. J. Time Ser. Anal. **1**, 15–30

17. Grenander, U., Szegö, G. (1956). *Toeplitz Forms and Their Applications* 2nd ed. University of California Press, Berkeley CA (Chelsea, New York, 1984)

18. Harville, D.A. (1997). *Matrix Algebra From a Statistician's Perspective.* Springer, New York

19. Hosking, J.R. (1981). Fractional differencing. Biometrika **68**, 165–176

20. Johnson, R.A. (1970). Asymptotic expansions associated with posterior distributions. Ann. Math. Stat. **41**, 851–864

21. Lawley, D. N. (1956). A general method for approximating to the distribution of likelihood ratio criteria. Biometrika **43**, 295–303

22. Lieberman, O., Rousseau, J., Zucker, D.M. (2000). Small-sample likelihood-based inference in the ARFIMA model. Economet. Theor. **16**, 231–248

23. Lieberman, O., Rousseau, J., Zucker, D.M. (2001). Valid Edgeworth expansion for the sample autocorrelation function under long range dependence. Economet. Theor. **17**, 257–275

24. Lieberman, O., Rousseau, J., Zucker, D.M. (2003). Valid asymptotic expansions for the maximum likelihood estimator of the parameter of a stationary, Gaussian, strongly dependent process. Ann. Stat. 31, to appear

25. Peers, H.W. (1965). On confidence points and Bayesian probability points in the case of several parameters. J. Roy. Stat. Soc. B **27**, 9–16

26. Philippe, A., Rousseau, J. (2000). Non informative priors in the case of Gaussian long-memory processes. Technical report, CREST, Paris

27. Robinson, P.M. (1994). Time series with strong dependence. In: *Advances in Econometrics*, Sixth World Congress (C. A. Sims, ed.) **1**, 47–95. Cambridge University Press, New York

28. Taniguchi, M. (1983). On the second order asymptotic efficiency of estimators of Gausian ARMA processes. Ann. Statist. **11**, 157–169

29. Taniguchi, M. (1990). Higher Order Asymptotic Theory for Time Series Analysis. *LNS* **68**. Springer, New York

30. Taniguchi, M., Kakizawa, Y. (2000). *Asymptotic Theory of Statistical Inference for Time Series*. Springer, New York

31. Welch, B., Peers, H.W. (1963). On formulae for confidence points based on intervals of weighted likelihoods. J. Roy. Stat. Soc. B **25**, 318–329

Contributors

Michael I. Baron, Programs in Mathematical Sciences, University of Texas at Dallas, Richardson, TX 75083-0688, USA

Bruno Bassan, Department of Mathematics, University of Rome 1, Piazzale Aldo Moro 5, I-00185 Rome, Italy

Martin Beibel, Research Group, Novartis Pharma AG, Klybeckstr. 141, CH-4057 Basel, Switzerland

Peter Dieterich, Institut für Physiologie, Technische Universität Dresden, Fetscherstraße 74, D-01307 Dresden, Germany

Adrian Dobra, National Institute of Statistical Sciences, Research Triangle Park, NC 27709-4006, USA

Stephen E. Fienberg, Department of Statistics and Center for Automated Learning and Discovery, Carnegie Mellon University, Pittsburgh, PA 15213-3890, USA

Konstantinos Fokianos, Department of Mathematics and Statistics, University of Cyprus, P.O. Box 20537, Nicosia 1678, Cyprus

Antonio Forcina, Department of Statistics, University of Perugia, Cp 1315, S1, 06100 Perugia, Italy

Seymour Geisser, School of Statistics, University of Minnesota, Minneapolis, MN 55455, USA

Nira Granott, Eliot-Pearson Department of Child Development, Tufts University, 105 College Avenue, Medford, MA 02155, USA

Joel L. Horowitz, Department of Economics, Northwestern University, Evanston, IL 60208, USA

Yuri I. Ingster, St. Petersburg State Transport University, Moskowskii av. 9, 190031 St. Petersburg, Russia

Márton Ispány, Institute of Mathematics and Informatics, University of Debrecen, Pf. 12, H-4010 Debrecen, Hungary

Athanassios Katsis, Department of Statistics and Actuarial Science, University of the Aegean, Samos, 83200, Greece

Jacob Kogan, Department of Mathematics and Statistics, University of Maryland and Baltimore County, Baltimore, MD 21250, USA

Zinoviy Landsman, Department of Statistics, University of Haifa, Mount Carmel, 31905 Haifa, Israel

Hans R. Lerche, Abteilung für Mathematische Stochastik, Universität Freiburg, D-79104 Freiburg, Germany

Offer Lieberman, Faculty of Industrial Engineering and Management, Technion-Israel Institute of Technology, 32000 Haifa, Israel

Charles F. Manski, Department of Economics and Institute for Policy Research, Northwestern University, Evanston, IL 60208, USA

Gusztáv Morvai, Department of Computer Science and Information Theory, Technical University of Budapest, 1521 Goldmann Grörgy tér 3, Budapest, Hungary

Gyula Pap, Institute of Mathematics and Informatics, University of Debrecen Pf. 12, H-4010 Debrecen, Hungary

Anne Philippe, Laboratoire de Mathematiques Appliquees, CNRS F.R.E. 2222, UFR de Mathematiques – Bat M2, Université des Sciences et Technologies de Lille, 59655 Villeneuve d'Ascq Cedex, France

Roland Preuss, Max-Planck-Institut für Plasmaphysik, Euratom Association, D-85748 Garching, Germany

Judith Rousseau, Laboratoire de statistiques, Université René Descartes, 45 rue des Sts peres, 75006 Paris cedex, France

Tamás Rudas, Institute of Sociology, Eötvös Loránd University, Budapest, Pázmány Péter sétány 1, 1117, Hungary

Andrew L. Rukhin, Department of Mathematics and Statistics, University of Maryland and Baltimore Country, Baltimore, MD 21250, USA

Marco Scarsini, Department of Statistics and Applied Mathematics, University of Turin, Piazza Arbarello 8, I-10122 Turin, Italy

Hans J. Schnittler, Institut für Physiologie, Technische Universität Dresden, Fetscherstraße 74, D-01307 Dresden, Germany

Albrecht Schwab, Institut für Physiologie, Universität Würzburg, Röntgenring 9, D-97070 Würzburg, Germany

Irina A. Suslina, St. Petersburg Institute for Exact Mechanics and Optics, Sablinskaya str. 14, 197101 St. Petersburg, Russia

Blaza Toman, National Institute of Standards and Technology, Statistical Engineering Division, Gaithersburg, MD 20899, USA

Emese Verdes, Department of Sociology, Debrecen University, Debrecen, Pf. 11, H-4010, Hungary

Shmuel Zamir, Center for Rationality, Hebrew University, Mount Scopus, 91905 Jerusalem, Israel

David M. Zucker, Department of Statistics, Hebrew University, Mount Scopus, 91905 Jerusalem, Israel

Martien C.A. van Zuijlen, Department of Mathematics, University of Nijmegen, Toernooiveld 1, 6525 ED Nijmegen, The Netherlands

Printing and Binding: Strauss GmbH, Mörlenbach